Programmed Introduction to Gas-Liquid Chromatography

Second Edition

J.B.PATTISON

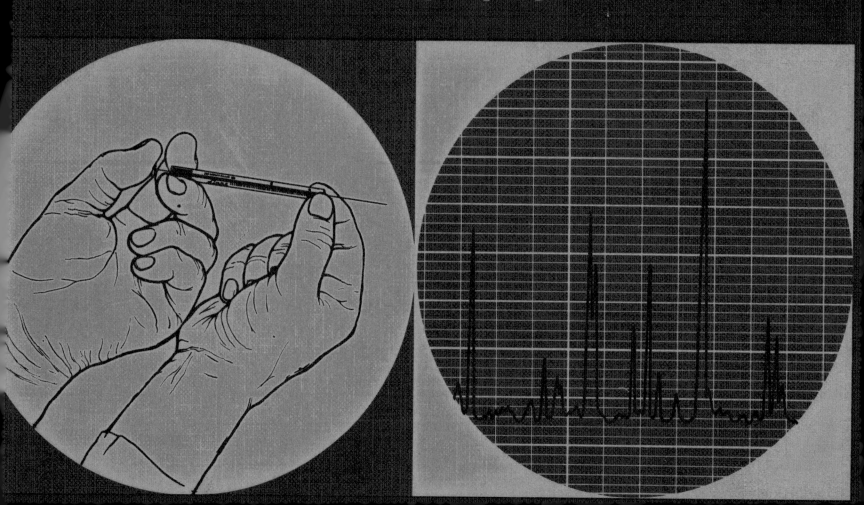

A Programmed
Introduction to
Gas–Liquid
Chromatography

A Programmed Introduction to Gas-Liquid Chromatography

by J. B. Pattison

Imperial Chemical Industries Limited
Petrochemical and Polymer Laboratory

SECOND EDITION

HEYDEN & SON LTD

London · New York · Rheine

Heyden & Son Ltd., Spectrum House, Alderton Crescent, London NW4 3XX
Heyden & Son Inc., 225 Park Avenue, New York, N.Y. 10017, U.S.A.
Heyden & Son GmbH, 4440 Rheine/Westf., Münsterstrasse 22, Germany

Library of Congress Catalog Card No. 69-19568
ISBN 085501 068 1 (paperback)
ISBN 085501 069 X (cloth)

Cover and layout by Peter Wright
Printed in Great Britain by Page Bros (Norwich) Ltd, Norwich

Contents

Biographical Note vi
Preface to the First Edition vii
Preface to the Second Edition viii
Objectives ix
Validation Report xi
The Criterion Test xiii

Teaching Programme

Part 1   Boiling Points and Vapour Pressure 1
     2   Partition Coefficients 17
     3   Chromatography 31
     4   Detection 52
     5   The Stationary Phase 78
     6   The Preparation and Packing of Columns 115
     7   Sample Injection and Syringe Technique 143
     8   Qualitative Analysis 155
     9   Quantitative Analysis 208
    10   Interpretation of the Chromatogram 229
    11   Conclusion 271

Summary and Exercises

Part 1, 2, 3 274
     4   275
     5, 6 276
     7, 8 277
     9   279
    10   281
Answers 282

Appendix A   Nomogram for Determination of the Retention Index   289
         B   Pressure Couplings 292
         C   A Simple Gas Chromatograph 293
         D   Syringes Currently Available 297
         E   A Guide to Further Reading 299
         F   Solid Support Equivalents Chart 301
Index 302

J. B. PATTISON was educated at A. J. Dawson Grammar School, Wingate, Co. Durham and the University of Sheffield. He joined ICI Ltd., (Research Dept.) Billingham, Co. Durham, in 1957 and worked on Exploratory Organic Chemistry. For some time, he was a member of the Heavy Organic Chemicals Division Research Department Safety Committee, where, together with three other members of the Committee, he produced a safety handbook and an associated safety film strip. Through introducing the film strip to students taking industrial safety courses, he became interested in modern methods of education.

In 1963 he transferred to the Petrochemical and Polymer Laboratory at Runcorn. During his career with the company, he has been involved not only in experimental work at the bench but also in the training of new laboratory assistants.

As a result of a short course on Programme Writing, led by John Hebenton (Training Superintendent, BP (Chemicals), Grangemouth, Stirlingshire) and arranged by Birkenhead Technical College, this programme was written.

GAS–LIQUID CHROMATOGRAPHY is now a common analytical technique in many chemical research laboratories, but to the new assistant with little or no knowledge of the subject, it may be a little confusing. Having been interested and involved in the training of new assistants, I felt that here was a subject in which the art of programmed learning might be of great use.

This book, which is the result of that idea, aims to provide a basic introduction to the subject which can be used as a foundation on which to build that comprehensive knowledge which must be gained from the standard texts and from practical experience in using the technique.

You work through the text at your own speed, taking an active part in the learning process. On every page you can prove to yourself that you have learned the material presented; usually a single fact or principle at a time. Having read the material on each page, you will be asked a question and invited to select the correct answer from several choices provided. Each choice is accompanied by a page reference and you turn to the page indicated for the answer you choose. If you choose the correct answer, the new page will confirm it and then go on to present new material. If you choose an incorrect answer, the new page will correct your mistake and you then go back to the question page to make another choice.

Ideally, you could go through the entire book without choosing a wrong answer, but there is no need to worry if you do make a mistake. Any mistake which you do make will be corrected immediately and you will learn the correct answer before tackling any new material.

The pages are numbered consecutively, but the information is not presented in sequence; for instance you may find the answer page before or after the question page. This is to prevent you from anticipating where the correct answer page may be found. **The important thing to remember is to work through the book strictly according to the instructions given on each page. However, if you should wish to take the argument backwards, you will find the page from which you came indicated at the top of each page on the side opposite the page number.**

To make the best use of the book, work in periods of about one hour, making notes as you go along. If possible, finish each period at the end of a distinct part. If you wish to revise any part before recommencing work, you will find summaries and exercises beginning on page 274.

The teaching of the programme should, if possible, be linked with experimental exercises using a g.l.c. machine.* As there are several different makes and types of machine in common use it has not been practicable to include these exercises in the programme itself, but no doubt the laboratory supervisor will cooperate at this stage.

You will need pencil, notebook and slide rule or tables of logarithms by your side as you tackle this programme, which starts on page 1.

I gratefully acknowledge the help and encouragement of my colleagues in the Petrochemical and Polymer Laboratory and of Mr. John Hebenton in the compilation of this programme. My thanks go too, to those in teaching establishments up and down the country who have helped in its validation.

June 1968.

Preface to the Second Edition

In the short time during which this programme has been in print many reports have been received of its effectiveness and usefulness in teaching the basics of Gas–Liquid Chromatography. The situations where it has been used have been many and various. I am very gratified with this favourable response.

Opportunity has been taken in this edition to make minor alterations and additions some of which have been drawn to my attention by users of the programme.

My thanks and appreciation go to all who have shown an interest in this work.

January 1973   J. B. PATTISON

---

* See *Appendix C* on page 293

Objectives

THE AIM is to teach the student the elementary general principles of gas–liquid chromatography, and then to apply those principles to the qualitative and quantitative analysis of organic liquid mixtures.

The student will be able:

1. To define the terms used in gas–liquid chromatography.

2. To select the appropriate stationary phase for any particular analytical problem.

3. To prepare his own columns.

4. To describe the function of the various units that compose a gas–liquid chromatograph.

5. To understand the selection of operating temperature, carrier gas flow and sample injection.

6. To interpret a chromatogram.

7. To analyse an organic liquid mixture qualitatively and identify the components.

8. To select the appropriate International g.l.c. Standard, and analyse an organic liquid mixture quantitatively.

9. To understand and use the Index of Kovats.

Validation Report

THE VALIDATION of this programme was based on a Criterion Test (which will be found on pages v–xiii). The test was constructed against the stated objectives, the outline of the test being compiled before the programme itself was written. This test was used both as a pre-test to eliminate students who already had too great a knowledge of the subject and as a post-test to measure the terminal success of the programme.

Students ranged from a majority of 'O' and 'A' level pupils in Grammar Schools, National Certificate students in Technical Colleges (some engaged in full-time courses, others working in industry and attending the Technical College on a part-time basis) and a few of graduate level. Their locations varied through Cheshire, Lancashire, Surrey, Yorkshire, Durham and Scotland.

All students first sat the test as a 'pre-test'. Those scoring more than 20% were eliminated on the grounds that they already had too great a knowledge of the subject to be suitable for evaluating the programme. Those scoring 20% or less in the pre-test then worked through the programme, study times being standardised at one hour per day. On completion each student re-sat the test, this time as a 'post-test'. No time limit was set for either the time to complete the programme or for writing the pre- and post-tests.

On first validation 29% of students achieved a score of 75% or higher and parts of the programme were rewritten in the light of the results obtained.

The second validation was carried out with a second group of students essentially similar in composition to the first. The same test was used as for the first validation and again any students scoring more than 20% in the pre-test were eliminated. Again study periods were controlled at one hour per day and unlimited time allowed for the pre- and post-tests. Validation with this rewritten programme gave 69% of students achieving 75% or higher marks in the post-test or 50% achieving 80% or higher.

The Criterion Test

1. Write down *seven* terms used in Gas Liquid Chromatography.
2. Define the term '*chromatography*'.
3. List *four* of the properties of a good stationary phase.
4. Which of the following stationary phases would you choose to separate a mixture of acetone (b.p. 56°) and methanol (b.p. 65°):
   (i) Apiezon L (non–polar); (ii) dinonyl phthalate (semi–polar); or (iii) Carbowax 20,000 (polar)?
5. List *five* precautions which should be taken in the preparation of a column.
6. Give the sequence of operations necessary when making an injection.
7. What is the proof of a successful injection?
8. Give three precautions to be taken when handling and storing a precision syringe.
9. Write down the formulae for *three* of the following:
   (i) Number of theoretical plates; (ii) Peak Resolution; (iii) Partition Coefficient; (iv) Peak Area Ratio; (v) Retention time.

Turn over for Question 10

10. Name the components of the g.l.c. apparatus shown in the block diagram below:

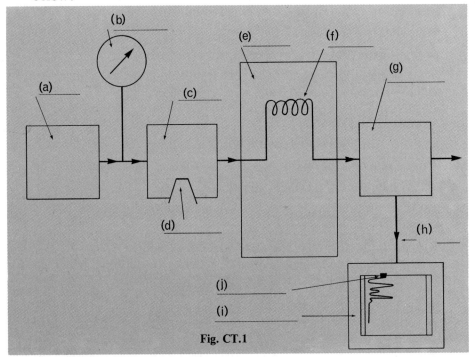

Fig. CT.1

11. Describe briefly what happens after a sample has been injected into the apparatus, concluding your account with the production of a visual trace.

12. Which part of a peak gives a measure of efficiency of the column?

13. Write down the two main classes of detectors and draw a sketch of the type of chromatogram produced by each.

14. (i) Which two types of detector are most commonly used? (ii) How does each type produce a signal?

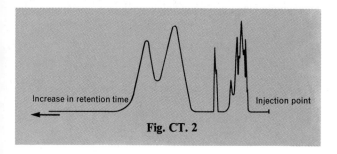

Increase in retention time

Injection point

Fig. CT. 2

15. Give a brief account of the way in which (i) Polarity and (ii) Hydrogen Bonding can help in a chromatographic analysis.

16. In a chromatographic analysis:
(i) What effect would a low temperature have on (a) efficiency and (b) retention times?
(ii) What would be the first choice of oven temperature?
(iii) Assuming the temperature were kept constant, how would a change in the flow of carrier gas affect the separation of two components?

17. How would you identify a component qualitatively by g.l.c.?

18. Which property of a peak gives a measure of the quantity of that component present in the sample?

19. What are the two factors which govern the choice of a suitable IUPAC International Standard?

20. In a quantitative analysis by g.l.c.:
(i) Give three of the four essential criteria for an internal standard.
(ii) What are the co-ordinates and abscissa of the calibration graph?
(iii) Between what limits is the calibration graph linear?

21. Give the Kovats Index for (i) methane; (ii) heptane; (iii) nonane.

22. A substance has a Kovats Index of 956 for a given stationary phase. Between which two alkanes will it be eluted if that phase was used for the analysis?

23. Look at the chromatogram opposite. Give three of the four possible ways of improving the resolution.

Now turn to page 285 and check your answers.

Part One

Boiling Points and Vapour Pressure

EVERYONE knows that water boils at 100°C and in so doing gives a vapour which we call steam. That this vapour exerts a pressure is shown every time a kettle boils. The pressure lifts the lid or blows a whistle!

Do you remember the relationship between the boiling point of a liquid and its vapour pressure? **If you do, turn to → page 11, if not read on here.**

If you have ever visited a hospital you will know that ether will vaporise at room temperature. It is a property of liquids, that they will evaporate and give a vapour.

Question
Which of the following statements is incorrect?

1. Alcohol at room temperature will evaporate and give a vapour → page 4

2. Mercury at room temperature will give a vapour → page 7

3. Iron at room temperature will give a vapour → page 10

B

Ether is a liquid, and obeys the law for liquids. At any given temperature, ether will have a definite vapour pressure. For instance you were told that at 35°C its vapour pressure was 760 mm. So then, suppose the vapour pressure of ether at room temperature were 600 mm. In the closed bottle, ether would evaporate until the pressure in the space above it was 600 mm, and then it would stop evaporating. As the stopper would be held on by the atmospheric pressure of 760 mm outside the bottle, a state of equilibrium would be reached.

Go back to → **page 10 and choose another alternative.**

back ref. page 5

We know from the question that at 35°C the vapour pressure of ether is 760 mm, and ether in an open vessel would boil off readily.

Would you expect the vapour pressure of tetralin (b.p. 207°C at 760 mm) to be low or high at 35°C?

In the light of your answer, whilst the ether was boiling off at 35°C would the vapour from the tetralin interfere or not?

Think about these two questions and when you can answer them, you should be able to answer correctly the question on → page 5.

Remember, the question asked for the incorrect statement! Alcohol is a liquid and obeys the laws for liquids, so at room temperature alcohol will evaporate and give a vapour.

Go back to ← page 1 and choose another alternative.

You have chosen the correct answer. Just as water in an open pressure cooker at 100°C and 760 mm pressure will boil and will eventually all be turned to steam, so ether at its boiling point of 35°C at 760 mm pressure will evaporate completely to give ether vapour.

If we hold a cold spoon in steam from the pressure cooker, we will condense some of that steam back to water again. On that simple fact depends the art of distillation, in which a mixture of liquids may be separated by condensing their vapours at their individual boiling points.

Question

Would you expect that a mixture of ether (b.p. 35°C) and tetralin (b.p. 207°C), (given that they do not react together),

1. ... would be impossible to separate by distillation? → page 9

2. ... would be easy to separate by distillation? → page 11

3. ... would be difficult to separate by distillation? ← page 3

You have chosen incorrectly. The fact that there are three liquids listed is not important, but the difference in the boiling points of the three liquids is important!

Remember, when the difference is great the separation is easy.

How many °C does the b.p. of benzene (80°C) differ from that of toluene (110°C)?

How many °C does the b.p. of toluene (110°C) differ from that of *cis*-cyclo-octene (138°C);

Look at your answers to these questions in the light of the statement above, then go back to → page 8 and choose the correct alternative.

Mercury is a liquid, and it obeys the laws for liquids, so at room temperature, mercury will give a vapour which, incidentally, is quite poisonous.

Go back to ← page 1 and choose another alternative.

Yes, the mixture of benzene (b.p. 80°C at 760 mm) and cyclo-hexane (b.p. 81·4°C at 760 mm) would be the most difficult to separate by distillation, because when the benzene boils at 80°C the vapour pressure of the cyclohexane would be almost 760 mm.*

In distillation it is the difference in boiling points that is important. *When the difference is great separation is easy, but when the difference is small separation is most difficult.*

Question
Now choose from the list below which you think would be the most difficult to separate by distillation. (All the boiling points are quoted for 760 mm pressure).

1. Benzene (b.p. 80°C), toluene (b.p. 110°C)   → page 12
2. Benzene (b.p. 80°C), toluene (b.p. 110°C), *p*-xylene (b.p. 138°C)   → page 14
3. Benzene (b.p. 80°C), toluene (b.p. 110°C), *cis*-cyclo–octene (b.p. 138°C)   ← page 6
4. *p*-Xylene (b.p. 138°C), *cis*-cyclo-octene (b.p. 138°C)   → page 16

---

* Their separation by g.l.c. is given on page 294.

At room temperature the vapour pressure of ether will be some-what below 760 mm. Let's say about 600 mm.

If we now raise the temperature of the mixture to 35°C whilst in an open vessel, will the ether boil off?

Will the tetralin (b.p. 207°C at 760 mm) boil off at the same time?

Think about these two questions and when you can answer them, you should be able to answer correctly the question on ← page 5.

Quite right. At room temperature iron does not give a vapour.

Now suppose we have some water in the bottom of a closed pressure cooker. At any given temperature, evaporation of the water will stop when the pressure inside the cooker builds up to a particular pressure of water vapour.

But, if the cooker were open and heat continuously applied to the bottom of the cooker, the water inside it would continue to evaporate until, at a particular temperature, the pressure of the vapour was equal to that of the atmospheric pressure. We would say then that the water was boiling, and if we looked into the open cooker, we would see bubbles of vapour forming in the water itself. We would call that temperature at which boiling occurred, the boiling point of the water. It is common knowledge that this is 100°C at atmospheric pressure that is 760 mm.

Gas–liquid chromatography depends partly on the fact that each liquid has its own particular boiling point.

Question
Given that ether boils at 35°C at 760 mm pressure, choose which of the following statements is correct:

1. Ether in an open beaker, at 35°C will evaporate
   completely. ← page 5
2. Ether almost filling a closed bottle at room
   temperature will evaporate completely. ← page 2

Yes, at the boiling point of a liquid, its vapour pressure is that of the atmosphere (usually about 760 mm). If a mixture of ether and tetralin were gradually heated up to 35°C (the boiling point of the ether) the ether would be completely vaporised, and this vapour could be condensed to yield pure ether. Then if the temperature were increased to 207°C (the boiling point of the tetralin), the remainder would be completely vaporised and this vapour could be condensed to yield pure tetralin.

This separation is easy because at the boiling point of ether (35°C) tetralin has a very low vapour pressure and so does not interfere in the distillation.

Question

Bearing the above statements in mind which do you think would be most difficult to do:

1. Separate ether (b.p. 35°C) and *m*-xylene (b.p. 138°C)?

→ page 15

2. Separate benzene (b.p. 80°C) and 2,2,4,-tri-methylpentane (b.p. 99·2°C)?

→ page 13

3. Separate a mixture of benzene (b.p. 80°C) and cyclohexane (b.p. 81·4°C)?

← page 8

At 80°C the vapour pressure of benzene would be 760 mm. At this temperature the vapour pressure of toluene (b.p. 110°C at 760 mm) would be lower than 760 mm.

Do you think the vapour pressure of toluene at 80°C would be low enough, to enable the mixture of benzene and toluene to be separated easily by distillation?

When you can answer this question, you should be able to answer correctly the question on ← page 8.

This separation would be difficult, but not the most difficult to do, so I am afraid you are wrong.

The two boiling points are very nearly the same but at 80°C the benzene will boil off and can be condensed.

At 80°C the vapour pressure of *iso*-octane (b.p. 99·2°C) will be quite high, almost 760 mm in fact, but there are distillation techniques which will allow the two vapours to be separated. The difference in boiling points is important.

When the difference is great the separation by distillation is easy.

Think what happens when the boiling points are very nearly the same. Go back to ← page 11 and choose the correct answer.

I am afraid you are wrong. Don't worry about the fact that alternative 2 has three liquids in the mixture. This is not important, but the difference in the boiling points of the three liquids is important.

Remember, when the difference is great then separation is easy. How many °C does the b.p. of benzene (80°C) differ from that of toluene (110°C)?

How many °C does the b.p. of toluene (110°C) differ from that of p-xylene (138°C)?

In the light of your answers and the statements above, go back to ← page 8 and choose the correct alternative.

You think that the mixture of ether (b.p. 35°C at 760 mm) and m-xylene (b.p. 138°C) would be the most difficult to separate. I am afraid you are wrong. The two boiling points are not nearly the same.

At 35°C the ether will boil off and can be condensed. At this temperature the vapour pressure of m-xylene will be very low and so separation will be easy.

It is the difference in boiling points that is important. When the difference is great then separation by distillation is easy.

Think what happens when the boiling points are nearly the same.

Go back to ← page 11 and choose the correct answer.

Yes, you have chosen correctly, for at 138°C both *p*-xylene and
cis-cyclo-octene will boil, and their vapour pressures will be
the same as that of the atmosphere, namely 760 mm.

In such a situation, however, gas–liquid chromatography would
be able to do the separation readily. To understand the process
of gas–liquid chromatography you need to know the theory of
'partition coefficients'.

If you already know this theoretical work, turn to → page 25.
If not, turn to Part 2 which begins on → page 17.

Part Two

Partition Coefficients

YOU WILL be familiar with a little bottle marked 'Tincture of Iodine' in the medicine chest. This simply is a solution of iodine in alcohol. Iodine will also dissolve in water and benzene. If a little iodine is added to a mixture of water and benzene, and the mixture is shaken until all the iodine dissolves, some of it will have dissolved in the water and some in the benzene.

If this is done at a constant temperature, then the concentration of iodine dissolved in the water (in g/ml) divided by the concentration of iodine dissolved in the benzene (in g/ml) is constant, no matter how much iodine we add. This constant is called the *partition coefficient*.

Question

If 3 g of iodine are completely dissolved in a mixture of 100 ml of water and 100 ml of benzene, it is found that 1 g of iodine dissolves in the water.

If 5 g of iodine are completely dissolved in a mixture of 100 ml of water and 100 ml of benzene, how much iodine would be dissolved in the water assuming that the temperature is the same in both examples?

1. 1g   → page 23

2. 1·67 g   → page 28

3. 3·3 g   → page 26

Wrong. Your answer shows that you have confused the quantities extracted from, and left dissolved in solvent A.

Remember it was calculated that at the first extraction, solvent B would extract 0·8 g of solute S from solvent A.

How much of solute S would be left in solvent A then?

Use your answer to calculate the amount of S that would be removed from solvent A at a second extraction with 100 ml of solvent B; then turn again to → page 28, and select the correct alternative.

You do not know whether it would be different or not? Suppose the number of grams of iodine per millilitre of water was C_1 and the number of grams of iodine per millilitre of benzene was C_2.

Then the partition coefficient $K_1 = \dfrac{C_1}{C_2}$

Now if we did the same for water and carbon disulphide we would obtain values of C_3 and C_4 say which would give $K_2 = \dfrac{C_3}{C_4}$

Even if $C_1 = C_3$, would it be possible for K_1 to equal K_2?

When you have thought of the answer, use it to choose the correct alternative on → page 22.

Your answer was none of those listed. Let me show you how to do it.

In the *first* extraction we have already seen that 0·8 g of solute S were extracted from solvent A. Since solvent A originally contained 1 g of S, this leaves $(1 - 0·8) = 0·2$ g of S in solvent A.

Let the amount of S extracted by solvent B at the second extraction be y g. How much of S does that leave dissolved in 100 ml of solvent A? The partition coefficient will still be $\frac{1}{4}$ and we know that

$$K = \frac{\text{no. of g of solute S dissolved in 1 ml. of solvent A}}{\text{no. of g of solute S dissolved in 1 ml. of solvent B}}$$

Substitute your values involving y into this equation, and solve it to obtain y; then go back to → page 28 and choose the correct alternative.

You say that there will be only one phase when water and benzene mix. Benzene can be obtained from crude oil by distillation and has oil–like characteristics.

If benzene and water were shaken together in a test tube, how many layers do you think there would be?

When you have thought of the answer turn again to → page 25 and choose another alternative.

Your answer 0·16 g is correct, for $K = \frac{1}{4} = \frac{(0·2 - y)}{y}$

Thus $5 y = 0·8$, so $y = 0·16$ g

If a third and fourth extraction were to be carried out, it would be found that 0·032 g and 0·0064 g of solute S would be removed from solvent A respectively.

We could represent the extraction process as shown in Fig. 2.1.

The application of these facts to gas–liquid chromatography will be seen in Part 3.

There are two important facts to learn about the partition co-efficient. (i) that the equation

$$K = \frac{\text{no. of g of solute dissolved in 1 ml of solvent A}}{\text{no. of g of solute dissolved in 1 ml of solvent B}}$$

only applies when the solutions are dilute. If the solubility of the solute in the solvents A and B is low, then we can add an excess of the solute and the equation will still hold.

(ii) that the numerical value of the constant known as the parti-tion coefficient depends on the solvent mixture used.

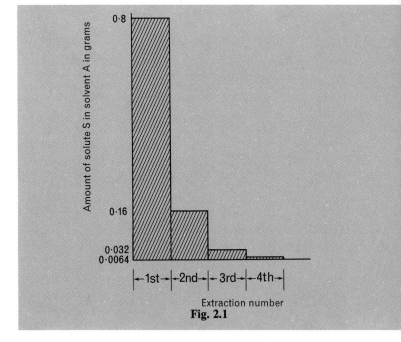

Fig. 2.1

Question

Iodine will dissolve in carbon disulphide as well as water and benzene. If a little iodine were dissolved in a mixture of water and carbon disulphide and the partition coefficient calculated, would its numerical value be different from that of iodine dissolved in a water/benzene mixture?

1. Yes.   → page 25

2. No.   → page 27

If you do not know   ← page 19

You think that 1 g of iodine would dissolve in the water. I'm sorry but you are wrong. Did you think that because 1 g dissolved in 100 ml of water in the first example, then 1 g would dissolve in 100 ml of water in the second example? This is not so.

The definition of *partition coefficient* can be re-written as an equation thus:

$$\text{partition coefficient} = \text{a constant} =$$

$$K = \frac{\text{no. of g of iodine dissolved in 1 ml of water}}{\text{no. of g of iodine dissolved in 1 ml of benzene}}$$

Thus the first step in solving the problem is to calculate K. To do this, using the values in the first example, calculate the number of grams of iodine dissolved in 1 ml of water. Now calculate the number of grams of iodine dissolved in 1 ml of benzene. Substitute these values in the equation to find K.

Now in the second example suppose x g of iodine dissolve in 100 ml of water, then $(5 - x)$ g of iodine are left to dissolve in the benzene. From these values calculate the number of grams of iodine dissolved in 1 ml of water. Now calculate the number of grams of iodine dissolved in 1 ml of benzene. Remember that we have already calculated K, so substitute the value for K and the values for the respective concentrations just found in terms of x, in the equation given above and you will have a simple equation to solve for x, which, remember, is the number of grams of iodine dissolved in 100 ml of water.

Compare your answer with those given on ← page 17 and choose the correct answer.

You have arrived at a wrong answer, which shows that you have confused the weights dissolved in solvents **A** and **B**.

After the first extraction we had 0·2 g of solute S dissolved in solvent A. Suppose in a second operation y g are extracted by solvent B. How much of S would there be left in solvent A? From the equation

$$K = \frac{\text{no. of g of solute S dissolved in 1 ml of solvent A}}{\text{no. of g of solute S dissolved in 1 ml of solvent B}}$$

recalculate y.

Turn again to → page 28 and select the correct alternative.

The partition coefficient of iodine in a water/benzene mixture would be quite different in numerical value from that of iodine in a water/carbon disulphide mixture, for *the numerical value of the constant known as the partition coefficient depends on the solvent mixture used.* Quite correct.

Although we have talked up to now about solutions of solids in liquids, *the same rules apply to solutions of gases and vapours in liquids or gases.*

It is well known that oil and water do not mix. We talk about 'pouring oil on troubled waters' and to this fact alone many sailors owe their lives.

There are many liquids which do not mix, and in all cases one floats on the other or others. Which floats on which depends upon the relative values of their specific gravities. Between immiscible liquids there may be one or several boundaries which divide them into distinct parts. We call each of these parts a PHASE, and in any mixture (or system) a phase is defined as a part of the system which is marked off from the other parts by a boundary, at which there is an abrupt change of physical properties (see Fig. 2.2).

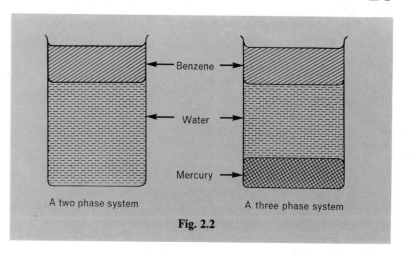

Fig. 2.2

Question

If some iodine were completely dissolved in a water/benzene mixture, how many phases would there be?

1. One. ← page 21

2. Two. → page 30

3. Three. → page 29

You have chosen the wrong answer.

We can write that the *partition coefficient* is a constant and is equal to

$$\frac{\text{no. of g of iodine dissolved in 1 ml of water}}{\text{no. of g of iodine dissolved in 1 ml of benzene}} = K$$

Now 1 g of iodine dissolves in the water in the first example, leaving 2 g dissolved in the benzene.

Calculate the concentration of the iodine in water.

Calculate the concentration of the iodine in benzene. Now substitute these values in the equation above to find a value for K, the constant known as the partition coefficient.

Now look at the second example. Suppose x g of iodine dissolved in the water, then $(5 - x)$ g would dissolve in the benzene.

Using these quantities calculate the concentration of iodine in water.

Now calculate the concentration of iodine in benzene.

Remember that we have already found a value for K, so if you now substitute your values in the equation at the top of the page you will have a simple equation to solve for x, the number of grams of iodine dissolved in the water.

Check your answer against those on ← page 17 and so choose the correct answer.

The numerical value of the partition coefficient changes if the solvent mixture is changed, so I am afraid you are wrong. The fact that the numerical value of the partition coefficient for the other solvent mixture is known, is no help here.

Go back to ← page 22 and choose another alternative.

Right. You have found that from the first example the partition coefficient for the system is 0·5. Then using this value in the second example you have found by simple algebra that 1·67 g of iodine would dissolve in 100 ml of water in the second example.

To apply partition coefficient theory to solvent extraction, let us suppose that 1 g of a substance S were dissolved in 100 ml of solvent A, 100 ml of solvent B (in which S is also soluble) were added, and the mixture was well shaken. Let the numerical value of the partition coefficient of S between solvents A and B be $\frac{1}{4}$. Then if the weight of S extracted by solvent B is x g, the weight of S remaining in solvent A is $(1 - x)$ g. We know that

$$K = \frac{\text{no. of g of solute S dissolved in 1 ml of solvent A}}{\text{no. of g of solute S dissolved in 1 ml of solvent B}}$$

or, $\frac{1}{4} = \dfrac{(1 - x)/(x)}{100 \big/ 100} = \dfrac{(1 - x)}{x}$

$\therefore x = 4 - 4x$

$x = \frac{4}{5} = 0·8$ g = the weight of S extracted by solvent B from solvent A.

Question
Now let solvent B be separated from solvent A, and the extraction repeated with a further 100 ml of solvent B.

On this second operation, how much of substance S will solvent B extract from solvent A?

1. 0·04 g   ← page 24

2. 0·16 g   ← page 22

3. 0·64 g   ← page 18

4. none of these amounts,   ← page 20

You say that there will be three phases in this system. This would be the case if the iodine had not completely dissolved in the solvent mixture.

In the light of this fact go back to ← page 25 and choose a better alternative.

We know that if some iodine were completely dissolved in a water/benzene mixture there would only be two phases, the iodine-in-water solution and the iodine-in-benzene solution, each solution being separated from the other by a definite boundary at which there is an abrupt change in physical properties.

We also know that there is a definite partition coefficient for this system. All that we have said for the system iodine/water/benzene applies to mercuric bromide/water/benzene, for mercuric bromide is soluble in both these solvents.

If, therefore, we took a mixture of iodine and mercuric bromide and added it to a mixture of water and benzene, then both iodine and mercuric bromide would dissolve in the two phases in a way controlled by their own particular partition coefficients as measured in separate experiments.

Since we have two distinct substances, namely iodine and mercuric bromide in our mixture, we say that the mixture has two COMPONENTS, both of which are soluble in our solvent mixture of two PHASES, water and benzene.

Now turn to Part 3 on page 31.

Part Three

Chromatography

WHEN THE components of a mixture are separated by making use of their differences in partition coefficients between two phases, (one of which should be mobile whilst the other is stationary) the term *chromatography* is applied. When a gaseous moving phase and a liquid stationary phase are used, the process is called *gas–liquid chromatography*. The liquid stationary phase is usually adsorbed on an inert solid which acts as a support, and the two phases come into contact in a tube known as a *column*. This column, usually of about $\frac{1}{8}$ to $\frac{3}{8}$ in. diameter, and made of metal, glass or plastic, is filled with the support material (which is impregnated with the stationary phase). Contact with the gaseous moving phase is obtained by passing that phase through the tube.

So then, to carry out a gas–liquid chromatographic analysis we need :

i) a tube to act as a container in which the phases can meet.

ii) a stationary phase.

iii) a gaseous moving phase, usually called the carrier gas.

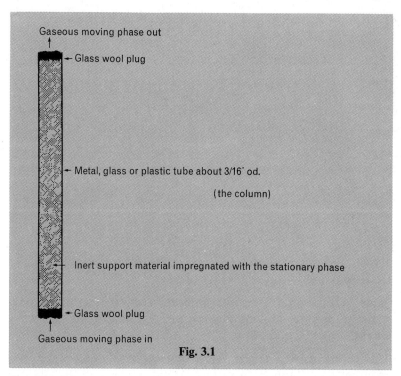

Gaseous moving phase out

Glass wool plug

Metal, glass or plastic tube about 3/16″ od.

(the column)

Inert support material impregnated with the stationary phase

Glass wool plug

Gaseous moving phase in

Fig. 3.1

Question

If we had a copper tube containing a liquid (such as liquid paraffin) supported on an inert solid like powdered firebrick, and we passed a gas, for example nitrogen, through it, do we have the basic conditions for gas–liquid chromatography?

1. Yes.　　　　　　　　　　　　　　　　　　　→ page 40

2. No.　　　　　　　　　　　　　　　　　　　　→ page 38

If you are not sure.　　　　　　　　　　　　　→ page 34

You don't know what would happen. Well let's work it out.

Regard our two sections as separate containers. Let us look at section 1 at equilibrium. Here it is:

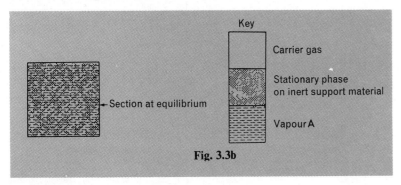

Key

Carrier gas

Stationary phase
on inert support material

←Section at equilibrium

Vapour A

Fig. 3.3b

It contains, in equilibrium, vapour A, gaseous moving phase (nitrogen) and liquid stationary phase.

Now at the same time, the adjacent section (section 2) only contains gaseous moving phase (nitrogen) and liquid stationary phase.

Let us now move the gas over from section 1 to section 2. Remember it contains both vapour A and nitrogen of the gaseous moving phase.

Since vapour A is soluble in the liquid stationary phase, some will dissolve in this phase in section 2, to an extent governed by the partition coefficient of the system.

Given these facts would you expect a new equilibrium to be set up in section 2?

Go back to → page 36 and choose the correct alternative.

You think that the retention time for vapour B would be some time other than x minutes. Quite right. Since the 'hopping' process is governed by the partition coefficient the retention volume for vapour B will change, say to V_1. Dividing this by the volume flow rate of the carrier gas (which you remember was kept constant) means that the retention time of vapour B must be different from x minutes. Let us say it was y minutes.

Suppose we now repeated the whole process described, keeping the conditions as before, but injecting a mixture of vapours A and B.

Question
Do you think that:

1. vapours A and B would be eluted together after $x + y$ min? → page 35

2. vapour A would be eluted after x min and vapour B after y min? → page 39

3. neither of these would occur? → page 50

You are not sure! Let me remind you that we set out to find the basic equipment for gas–liquid chromatography. Let me try to allay your doubts with an analogy.

If you wanted to cook some potatoes, one way that you could do it would be to put some peeled potatoes into a pan containing water which on being heated would boil and in so doing would cook the potatoes.

The basic conditions which you need to do this are:

(i) peeled potatoes

(ii) water

(iii) a pan to contain them both

(iv) some means of heating the water.

In other words two 'phases' meet in a container, which is heated externally.

From this analogy go back to ← page 31 and choose another alternative.

Your answer, that vapours A and B would be eluted together after $x + y$ minutes is wrong.

Each vapour, in the mixture, would behave only in accordance with its own particular partition coefficient during the chromatographic process. Thus their retention times when mixed together would be the same as for the individual components, provided all other conditions remained constant. When analysed separately vapour A's retention time was x min and vapour B's retention time y min.

Go back now to ← page 33 and choose the correct alternative.

Right. You have remembered that the theory of partition co-efficients applies to solutions of gases and vapours in liquids or gases. The small quantity of vapour A would dissolve partly in the stationary phase and partly in the gaseous phase to an extent dependent on the partition coefficient for this system, so that a vapour equilibrium with the gaseous phase is established (Fig. 3.3). This state of affairs would only be possible if the flow of the gaseous moving phase ceased, but in fact the flow is continuous.

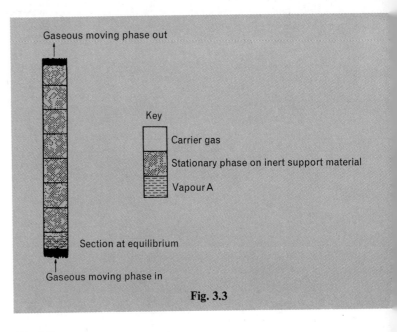

Gaseous moving phase out

Key

Carrier gas

Stationary phase on inert support material

Vapour A

Section at equilibrium

Gaseous moving phase in

Fig. 3.3

Question

Imagine now that the volume of gas and vapour A in the section moves on, in a 'block' as it were, to the next adjacent section and is replaced by pure gaseous moving phase. Would you say that in that next section:

1. nothing would happen? → page 4

2. a new equilibrium would be established between the two phases and vapour A? → page 4

If you do not know what would happen. ← page 3

Wrong. The volume V called the retention volume depends on the solubility of the solute in the stationary phase. This volume is low for solutes which are slightly soluble in the stationary phase and high for solutes which are very soluble in the stationary phase.

Does the retention volume change if the value of the partition coefficient changes?

Since the volume flow rate of the carrier gas in our example remains constant, what can you then say about the retention time of vapour B?

When you have answered these questions correctly you will be able to choose the correct alternative on → page 45.

I'm afraid you are wrong!

Let us examine the question again remembering the three parts of the definition of *chromatography*.

(i) Have we a gaseous moving phase?

(ii) Have we a stationary phase?

(iii) Have we a tube containing the stationary phase, and in which the gaseous moving phase can meet it?

In the light of your answers, go back to ← page 31 and choose another alternative.

Correct. The behaviour of the components of the vapour mixtures would be governed only by their own particular partition coefficients, so each component would behave exactly as it did when it alone was mixed with the carrier gas, and in this case, vapour A would have a retention time of x minutes and vapour B of y minutes.

You remember that we regarded our column as made up of distinct equal sections or 'plates'. You will learn later in the programme how to calculate how many plates a column contains, in theory, for any separation. The *efficiency* of the column is related to this number of plates, and a measure of this efficiency is the width of the peak. The *narrower* the peak the *higher* the efficiency and vice versa.

With most solutes, a change in temperature will affect their solubility in the solvent. This would alter the partition coefficient. The *more* solvent there is present as stationary phase, the *more* solute will dissolve in it. The effect of these facts in a g.l.c. analysis would be to alter the retention volume, and hence the retention time, even if the volume flow rate of the carrier gas were to be held constant.

Question

A vapour C was chromatographed on a suitable column at a temperature $x°$ using nitrogen as carrier gas at a volume flow rate of v cm^3/sec. Its retention time was found to be t sec.

The analysis was repeated using the same column at $x°$ with nitrogen as carrier gas at a volume flow rate of w cm^3/sec ($v \neq w$). Would you expect the retention time in the second analysis to be:

1. t sec? → page 43

2. some time other than t sec? → page 51

If you do not know → page 48

You are quite correct. We have a container—the copper tube; a stationary phase—liquid paraffin on powdered firebrick; a gaseous moving phase—nitrogen, so we have the basic conditions for gas–liquid chromatography.

Let us look at our column again (Fig. 3.2a). Although the column is continuous, for theoretical purposes we could regard it as being composed of distinct equal sections or 'plates' as they are called (Fig. 3.2b). Now each of these sections can be considered separately and so in each we have a two phase system, consisting of a gaseous moving phase and a liquid stationary phase supported on powdered firebrick.

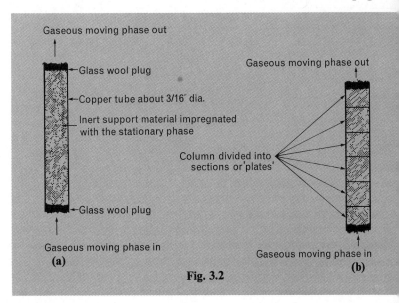

Gaseous moving phase out

Glass wool plug

Copper tube about 3/16″ dia.

Inert support material impregnated with the stationary phase

Glass wool plug

Gaseous moving phase in

(a)

Gaseous moving phase out

Column divided into sections or 'plates'

Gaseous moving phase in

(b)

Fig. 3.2

Question

Imagine just one of these sections by itself. If a small quantity of a vapour A which was soluble in the stationary phase were introduced into the section, it would diffuse throughout that section, mixing completely with the gaseous moving phase.

In these circumstances, do you think that partition would take place between the two phases and vapour A?

1. Yes ← page 36

2. No → page 46

If you are not sure. → page 44

You are not sure what the retention time of vapour B would be, well let us look again at the facts. We know that:

(i)   the retention volume V divided by the volume flow rate of the carrier gas gives us the retention time of any component.

(ii) the volume flow rate of the carrier gas has remained constant.

(iii) the partition coefficient of vapour A in the two phases is different from that of vapour B in the two phases. Let us say they are K_A and K_B.

(iv) Because K_A is not equal to K_B the solubilities of vapour A and vapour B in the stationary phase will *not* be the same.

That being the case, and knowing that the retention time of vapour A was x minutes, what can you now say about the retention time of vapour B?

Turn again to → page 45 and choose the correct alternative.

Correct. A new equilibrium would be established between the vapour carried into it by the carrier gas (i.e. the gaseous moving phase), the carrier gas itself and the stationary phase in that second section. The partition coefficient does not change since the two phases have not changed and the solute, vapour A, is the same

Meanwhile, in the first section, a new equilibrium will be established too. Vapour A is removed from solution in the stationary phase in that first section, by pure carrier gas which will have entered it from the gas supply.

Although we have imagined the process taking place in steps, in fact it is a continuous process, and vapour A will be carried by this 'hopping' process, through the column.

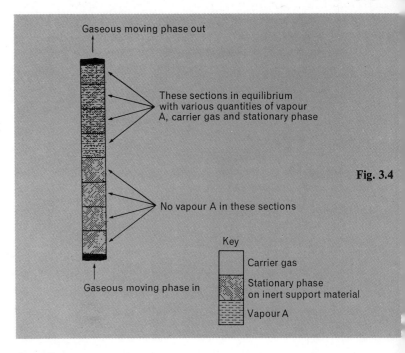

Fig. 3.4

Question

Now consider what happens at the exit end of the tube. Let us look at the last few sections, after carrier gas has been passing at a constant volume flow rate for a while (Fig. 3.4).

If another 'hop' occurred would you expect that:

1. only carrier gas would come out at the top of the column? → page 49

or

2. carrier gas and vapour A would come out at the top of the column? → page 45

You would expect the retention time in the second analysis to be t secs. I'm afraid you are wrong.

Although the column temperature has been kept constant at $x°$, have you noticed that the volume flow rate of the carrier gas has changed from v cm^3/sec to w cm^3/sec? What effect would this velocity change have on the retention time of vapour C?

Return to ← page 39 and choose the correct alternative.

You're not sure whether partition would take place between the two phases. Well let's look again at the facts.

(i)   The theory of partition coefficients applies to solutions of gases or vapours in liquids or gases.

(ii) From the gas laws, vapour A will mix with the nitrogen of the gaseous moving phase.

(iii) You have been told that vapour A is soluble in the liquid stationary phase.

Now decide whether these facts confirm that partition would occur between the two phases in our chosen section, then return to ← page 40 and choose the correct alternative.

back ref. page 42   45

Yes, carrier gas and vapour A will come out at the top of the column. As the process continues the amount of vapour A (due to the extraction process described) would gradually increase to a maximum then fall away to zero when all of it had passed through the column. We could represent this graphically (see Fig. 3.5). This curve is called the *'elution' curve*.

Point M on the curve represents the maximum of this curve. To reach this point on the curve a definite volume of carrier gas will have passed through the column. This volume V, called the *retention volume* will be low for solutes which are volatile or less soluble in the stationary phase (i.e. value of K low) and high for nvolatile solutes or those which are very soluble in the stationary phase (i.e. value of K high). When the retention volume is divided by the volume flow rate of the carrier gas, the *'retention time'* is obtained.

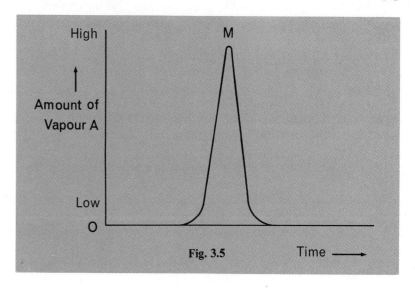

Fig. 3.5

Question

Suppose that we now repeated the whole process described using a different vapour B whose partition coefficient in the two phases was different from that of vapour A in those phases. Given that the retention time for vapour A was x minutes and the other conditions have not changed would you expect that the retention time for vapour B would be:

1. x minutes   ← page 37

2. some time other than x minutes   ← page 33

If you are not sure.   ← page 41

Your answer, that partition would not occur, is wrong.

Remember all that was said regarding the theory of partition coefficients applies to solutions of gases or vapours in liquids or gases.

From your knowledge of the gas laws would vapour A and the nitrogen of the gaseous moving phase, mix together?

Would some of vapour A dissolve in the liquid stationary phase?

When you have answered these questions, return to ← page 40 and choose the correct alternative.

You think that nothing would happen. I'm afraid you are wrong.

Let us look at our two sections again before the gas moves over. Here they are.

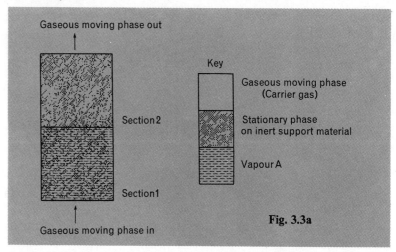

Fig. 3.3a

Section 1 contains, in equilibrium: vapour A, gaseous moving phase (nitrogen) and stationary liquid phase; whilst section 2 contains only gaseous moving phase (nitrogen) and stationary liquid phase.

Now let us imagine that the gas contained in section 1 moves over in a 'block' to section 2. Remember it contains both vapour A and nitrogen of the gaseous moving phase, and since vapour A is soluble in the liquid stationary phase, some of it will then be extracted from the gaseous phase by the liquid stationary phase, to an extent governed by the partition coefficient of the system. Given these facts would you expect a new equilibrium to be set up in section 2?

Return to ← page 36 and choose the correct alternative.

back ref. page 39

You don't know what the retention time would be? Well let's work it out.

We know that:

(i) retention volume changes if the column temperature changes. But in both analyses the column temperature was constant at $x°$. Therefore the retention volume will be the same in both cases.

(ii) retention time $= \dfrac{\text{retention volume}}{\text{volume flow rate of the carrier gas}}$

Now let us say the retention volume was V cm^3

From the first analysis we have

$$t = \frac{V}{v} \qquad\qquad (1)$$

Let the retention time in the second analysis be t' sec. We then have

$$t' = \frac{V}{w} \qquad\qquad (2)$$

From equations (1) and (2) in which circumstances would t' be equal to t?

Return and choose a better alternative from those on ← page 39.

I'm afraid you are wrong. As the carrier gas moves along the tube it will be in equilibrium with the liquid stationary phase and vapour A. At the next 'hop', the volume of gas in the last section, composed of a solution of vapour A in carrier gas, will be emitted from the column.

Return to ← page 42 and choose the other alternative.

You think that the retention times would be neither **1.** Vapour A at x min and vapour B at y min, nor **2.** Vapours A and B together at $(x + y)$ min. Fine, I'm glad you're honest about it, so I'll try to make things clear with an analogy.

Imagine two trains standing at a London Terminus, one an express, the other a slow train going to the same destination. Both are due to leave the terminus at the same time say 4.30 p.m. When that time arrives, both commence to leave the terminus, but the express picks up speed faster and so sets off down the track ahead of the slow train. On they go, the express first, the slow train second, but *both on the same track.*

The express will reach the destination ahead of the slow train, say at 6.30 p.m., but when it does, the slow train will still be chugging along the track at its own speed. Finally, the slow train will arrive at the destination, say at 7.00 p.m., having taken a longer time to do so.

Using the above analogy, if one train were called 'Vapour A' and the other 'Vapour B', and the track were the path through our column, go back to ← page 33 and choose the correct alternative from those listed there.

Quite right. The retention volume would be the same in both cases. Let us call it V cm^3. To obtain the retention time this volume must be divided by the volume flow rate of the carrier gas.

In the first case the retention time $t = \dfrac{V}{v}$ sec.

In the second case it is $\dfrac{V}{w}$ sec. Since $v \neq w$ this is clearly different from t sec.

We can summarize the factors which affect *retention time* as follows:

(i)   Change in column temperature.

(ii) The amount of stationary phase impregnated on the support material.

(iii) The volume flow rate of the carrier gas.

(iv) The length of the column.

Part 4 starts on → page 52.

Part Four
Detection

WE HAVE seen that it is possible to separate the components of a mixture by gas–liquid chromatography, but we still have the problem of detecting these components. We could use a micro condenser, watch the droplets appear, and measure the time taken for them to appear, but this would be too slow and too inaccurate for our needs.

To solve the problem, several different types of electrical detector have been devised.

Turn now to → page 70, read the narrative contained in it, then without reference to it, answer the following questions.

1. Which are the two main classes of detectors?
2. To what is the signal proportional in a mass-sensitive detector?
3. What is a katharometer?
4. How does it work?
5. In which type of detector is a thermocouple used to produce the signal?
6. Describe briefly the operation of the gas density balance type of detector.
7. Which detector depends on the fact that combustion of the components in a hydrogen flame causes ionization?
8. What three criteria do the best detectors possess?

Now turn to → page 283 and check your answers against those given there.

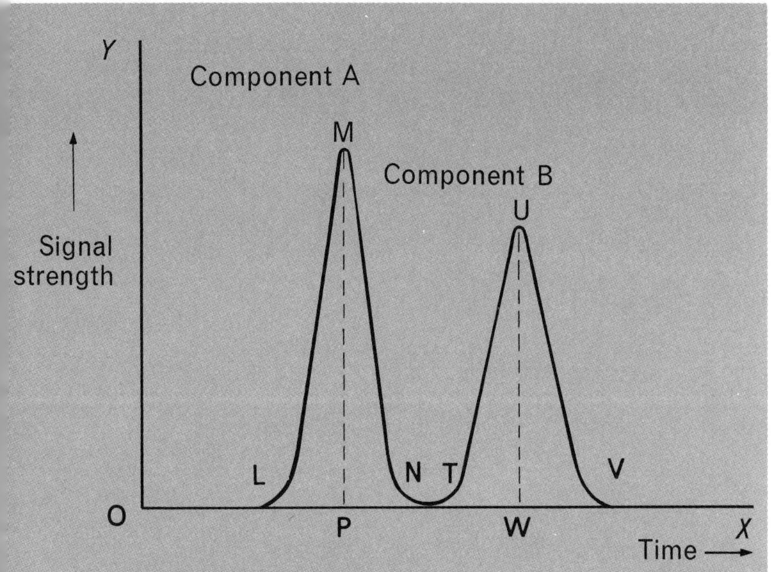

Fig. 4.5c

You're not sure how much of each component there was in the original mixture. Let's see how to calculate it. Do you remember that we said (→ page 68) that the *total quantity of each component which has passed through the detector is represented by the area under its peak in the chromatogram.* Good. Let us look at our chromatogram again (Fig. 4.5c).

The first thing we need to do is to calculate the *peak areas* for each component.

Component A: Peak Area LMN = $\frac{1}{2}$ (LN) (PM)

Component B: Peak Area TUV = $\frac{1}{2}$ (TV) (WU)

Now, you were told that LN = 20 mm; PM = 35 mm; TV = 28 mm and WU = 25 mm.

Calculate areas LMN and TUV from this information.

Are these areas equal?

What can you then say, (remembering the relationship between *quantity* and *peak area*) about the amounts of components A and B in the original mixture?

Return to → page 63 and choose the correct alternative.

Using the results given, you have substituted them in the first formula

$$n = 16 \left(\frac{t}{W_b}\right)^2$$

and found that n = the number of theoretical plates = $16 \left(\frac{56}{7}\right)^2$

$$= 16 \times 64 = 1{,}024 \text{ plates.}$$

Then from the formula $H = \dfrac{\text{length of the column}}{n}$

by substituting the length of the column in millimetres and the value for n just found, you have calculated that

$$H = \frac{2{,}000}{1{,}024} = 1{\cdot}95 \text{ mm.}$$

Quite correct.

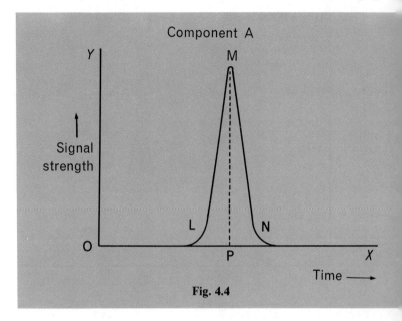

Fig. 4.4

Question

Let us look at the chromatogram again. What else can we ge
from it? Do you think we can calculate the *total amount* of the
component A which has passed through the detector by
measuring,

1. the peak height MP? → **page 6**

2. the peak width LN? → **page 6**

3. the peak area LMN? → **page 68**

ou have calculated that each area LMN and TUV is equal to
50 mm². Then since the areas are equal, there must have been
qual amounts of the two components A and B in the original
ixture. Right.

lthough the areas are the same, have you noticed that the
eights and widths of the peaks are different? *The width of a
eak depends only on the chromatographic process and not on the
etector used.* Now let us see what else we can get from our
hromatogram (Fig. 4.6). If we draw in the inflection tangents
R, QS; and JY, JZ, for each peak, then the *Peak Resolution* (or
eparation Factor) R, is defined as:

$$\frac{\text{(distance from one max. peak height to the other)}}{\text{sum of the base widths of each peak}} = \frac{2(PW)}{RS + YZ}$$

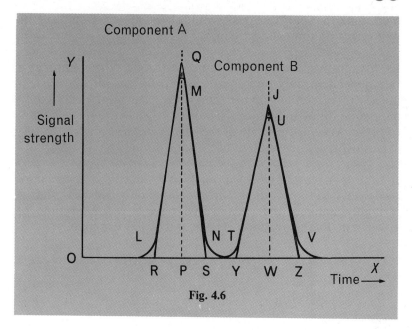

Fig. 4.6

Question
Given that the distance between the maximum peak heights for
components A and B is 48 mm, the width of the base of peak A
is 12 mm and the width of the base of peak B is 20 mm, the
peak resolution is calculated as:

1. 3 → page 74

2. 1·3 → page 76

3. neither of these. → page 67

back ref. page 68

You think that only one peak would show on the chromatogram at a retention time of $x + y$ min. This is wrong.

Each vapour in the mixture would go through the column at a rate dependent only on its partition coefficient for that system unless (i) the column temperature were changed (which would alter solubilities and hence the value of the partition coefficient) or (ii) the volume flow rate of the carrier gas was altered. In this experiment, neither of these two events occurs.

If vapour A reacted with vapour B to produce a different pure compound vapour C then there *would* be only one peak, but it is highly unlikely that this would have a retention time of $x + y$ min. Such a reaction is ruled out in this example.

In view of these conclusions, go back to → page 68 and choose a better alternative.

Your answer—change the detector—is wrong!

If the detector is changed the only factor that changes is the *sensitivity* of detection of the components.

It has been stated (← page 55) that *the width of a peak depends only on the chromatographic process and not on the detector used.*

Does the width of the peaks affect the **Peak Resolution, *R*?**

Think out the answer to this question then go back to → page 74 and choose the correct alternative.

Each differential detector produces an electrical signal proportional to the quantity of each component present, so the signal strength will rise from zero to a maximum, *M*, then fall back to zero as each component is eluted. *M* corresponds to the *retention volume* and the time taken to reach *M*, measured on the *OX* axis of the graph, is the *retention time*, provided the volume flow rate of the carrier gas and the temperature are kept constant. Fig. 4.1 shows this graph, which is called a *chromatogram* for one component. This *chromatogram* contains one *peak*.

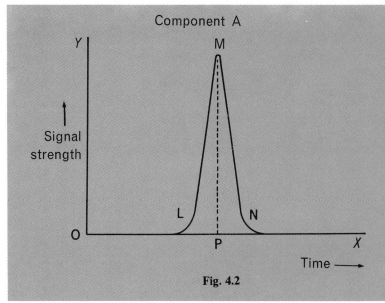

Fig. 4.2

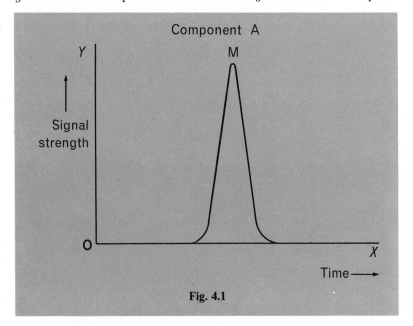

Fig. 4.1

The horizontal line through the point of zero signal strength indicates that no component was present in the detector. This line is called the *base line*.

Question

Now look at Fig. 4.2. If component A was introduced on to the column at 0 (the *injection point*), would you say that the retention time for component A was:

1. OL → page 60

2. OP → page 66

3. LN → page 64

provided the volume flow rate of the carrier gas remained constant?

Your answer, 0·51 mm, is wrong. You have calculated the number of theoretical plates correctly as

$$n = 16 \left(\frac{56}{7}\right)^2 = 16\,(8)^2 = (16)\,(64) = 1{,}024.$$

Now $H = \dfrac{\text{length of column}}{1{,}024}$

You know that the length of the column is 2,000 mm, so calculate the value of H.

Now return to → page 66 and select the correct alternative.

I'm afraid you are wrong. As long as the signal remains at zero strength, a horizontal line through zero will appear on the chromatogram.

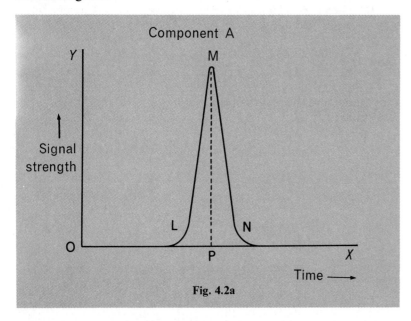

Fig. 4.2a

As soon as the signal strength begins to rise (as it does at L) we know that some of component A is beginning to pass through the detector. In view of these facts, what does OL represent?

Return now to ← page 58 and choose a better alternative.

No. The signal strength at any time is proportional to the quantity of the component passing through the detector at that time.

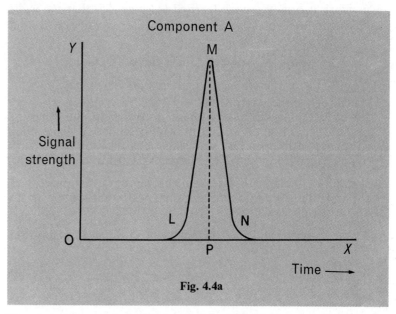

Fig. 4.4a

The signal strength indicated by the height MP represents only that amount passing through the detector at a time after injection, given by the length OP.

Return to ← page 54 and choose another alternative.

You have no idea what to do! Well let us look at the other alternatives in turn.

(i) *Change the detector*—would this improve the separation? Let us work it out. We have learned (← page 55) that *the width of a peak depends only on the chromatographic process and not on the detector used.*

   Does the width of the peaks affect the peak resolution R?

(ii) *Use a column containing a different stationary phase*—would this improve the separation?
   If we change the stationary phase, do you think the solubility of our two components in it will be different from their solubility in the original stationary phase? Would a change in solubility bring about a change in partition coefficient K?

   Would a change in K alter the chromatographic process? If the chromatographic process changes, what can you say about the peak widths?

   Again, does the width of the peaks affect the peak resolution, R?

In the light of your answers to these questions, return to → page 74 and choose the better alternative.

Correct. There would be two peaks, the retention times of which were x min and y min, the same as for each component on its own; if all the other conditions are kept constant, retention volume and hence retention time are affected only by the partition coefficient. The chromatogram could look something like Fig. 4.5.

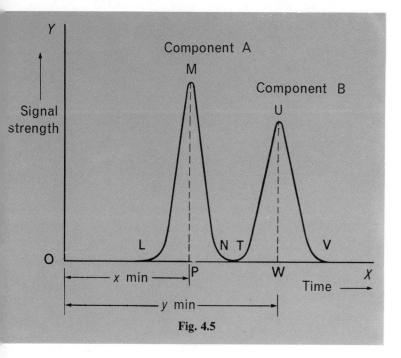

Fig. 4.5

Question

For component A, (say heptane), the maximum peak height PM is 35 mm, and the width at the base LN is 20 mm.

For component B, (say octane), the maximum peak height UW is 25 mm, and the width at the base TV is 28 mm.*

Assuming that the peaks are approximately triangular in shape, calculate the areas LMN and TUV from this information. From these areas can you say that in the original mixture there was

1. an equal amount of component A and component B?
← **page 55**

2. more of component A than component B?   → **page 72**

3. more of component B than component A?   → **page 75**

If you are not sure how much of each component there was in the original mixture turn to ← page 53.

---

* To within experimental error, the response factors for heptane and octane are equal.

You say that the retention time for component A is LN. You are wrong.

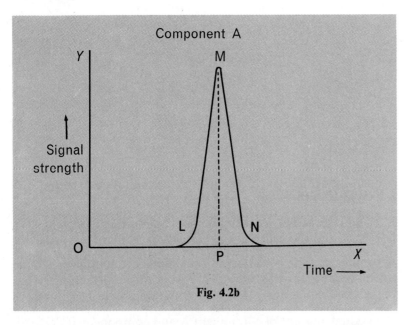

Fig. 4.2b

L marks the time when some of component A is beginning to pass through the detector. N marks the time when, once again, the signal strength has returned to zero, signifying the absence of component A in the detector.

Using these facts, what does the time LN represent?

Return to ← page 58 and choose a better alternative.

No. The width of the peak LN represents the time taken for the component to pass through the detector, not the quantity that has passed through.

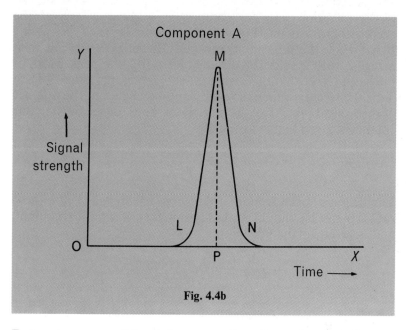

Fig. 4.4b

Return to ← page 54 and choose another alternative.

You are quite right. Since the signal is proportional to the quantity of the component present at any time in the detector, M, the maximum point, corresponds to the retention volume, and since the volume flow rate of the carrier gas is constant, OP is the retention time for component A. Let us look at the chromatogram again. (Fig. 4.3).

If we draw the inflection tangents QR and QS we can then measure with a ruler the distance RS = W_b (the peak width at the base). (Note that this distance really represents a time!) Then if t is the retention time measured in the same units, the number of theoretical plates (i.e. the number of 'hops' or sections) in our column is given by the formula:

$$n = 16\left(\frac{t}{W_b}\right)^2$$

The height of each section of the column, equivalent to a theoretical plate (H), should be about 1 mm for a well packed column, and is given by the formula:

$$H = \frac{\text{length of column}}{n}$$

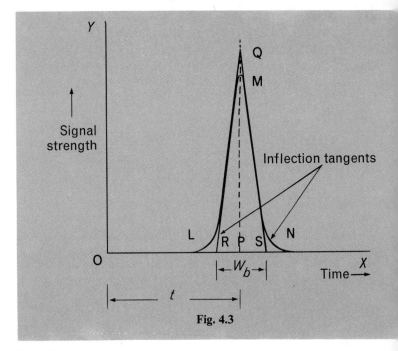

Fig. 4.3

Question

In an experiment to find the height equivalent to a theoretical plate (H) for a particular 2 metre column, the following results were found after drawing the inflection tangents to the peak obtained:

(i) peak width at the base 7 mm, (ii) retention time 56 mm.

From this information would you say that the value of H for this column was

1. 0·51 mm?
2. 1·95 mm?

If you cannot work this out.

← page 5
← page 5
→ page 6

So your answer agrees with neither of the alternatives given!
I think we had better work it out step by step. Here is the
chromatogram again.

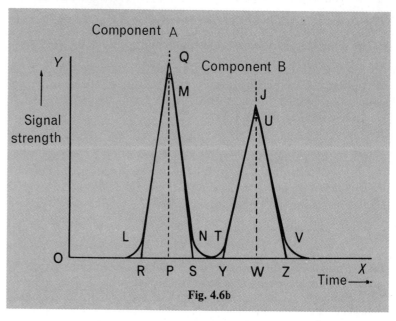

Fig. 4.6b

The Peak Resolution $R = \dfrac{2(PW)}{RS + YZ}$

Now PW was given as 48 mm; RS as 12 mm and YZ as 20 mm.
By substitution we have

$$R = \frac{2(48)}{12 + 20} = \frac{2(48)}{32} = \;?$$

**Finish the calculation to find R, then choose the correct alterna-
tive on ← page 55.**

Quite right. The quantity of any component which has passed through the detector is given by the equation:

Total Quantity = Peak Area × Detector Response Factor.

The numerical value of the response factor depends on:

(i) the detector, and (ii) the substance being detected.

As long as we are considering one particular detector and one particular substance, then the total quantity of that component which has passed through the detector will be represented by the area under the peak LMN.

For the most accurate work, this area is measured by a planimeter or by using an integrator which is fitted to the recorder, but for most practical purposes, we can assume that the peak is approximately the same shape as a triangle. The area LMN is then approximately equal to

$\frac{1}{2}$ (base) × (height)
$= \frac{1}{2}$ (LN) × (MP) = Total quantity of component A.

The applications of this will be seen when we find out later how to do quantitative analysis by g.l.c.

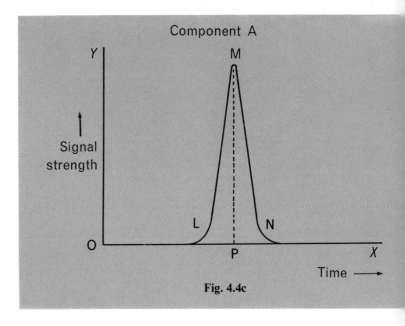

Fig. 4.4c

Question
Suppose we now inject on to the column a mixture of vapour A and B which do not react with one another and the retention times of which, when injected separately, are x minutes and y minutes respectively. If all other conditions are kept constant would you expect the chromatogram to show:

1. two peaks, the retention times of which were **not** x min. and y min? → page 7

2. two peaks, the retention times of which **were** x min. and y min? ← page 6

3. one peak the retention time of which was $x + y$ min? ← page 5

You cannot work out the value of H. Let me show you how to do it. Firstly the formula for H is

$$H = \frac{\text{length of column}}{n} \qquad (1)$$

where n is the number of theoretical plates in the column and is given by the formula

$$n = 16\left(\frac{\text{retention time}}{\text{peak width at its base}}\right)^2 = 16\left(\frac{t}{W_b}\right)^2 \qquad (2)$$

Secondly the retention time was 56 mm and the peak width at its base was 7 mm, so you can substitute these values in Equation (2) and find n. You were also told that the column length was 2 metres (i.e. 2,000 mm). Substitute this value and your calculated value of n in Equation (1) and you will have found the value of H for the column.

Check your answer against those on ← page 66 and select the correct alternative.

'Detectors'*

As the components leave the column one after the other they are quantitatively analysed by the detector. The best detectors are sensitive, reliable and accurate. At present there are two detection systems.

(1) DIFFERENTIAL DETECTORS which give a signal, the magnitude of which is proportional to the amount of the component in the detector (concentration-sensitive detector) *or* to the mass of the component which passes through the detector per unit time (mass-sensitive detector). Some important differential detectors are:—

(a) The *thermal conductivity cell* or *katharometer* which consists of a thin wire which is part of a Wheatstone bridge circuit. The wire is heated electrically. As the gas mixture passes over the wire there are variations in its thermal conductivity which cause variations in the equilibrium temperature of the wire and consequently its resistance. These changes can be used to give a chromatogram.

(b) *The flame ionization detector.* Combustion of the solute components in the carrier gas in a hydrogen flame causes ionization. The conducting gases cause a current to flow between two electrodes which are held at constant potential. This current is used to provide the chromatogram.

(These two types of detector are the most commonly used types).

(c) *The gas density balance.* In this detector the density of the gas mixture leaving the column is continually compared with that of pure carrier gas.

(d) *The flame detector.* The hydrogen used as carrier gas is burned at the end of the column. As the solute components leave the column, they burn, the flame lengthens and its temperature increases. The temperature, measured by a thermocouple, is used to provide the chromatogram.

(e) *The ionization detector.* The solute components in the carrier gas are ionized by β-rays discharged from a radioactive source (e.g. Sr^{90}). An ionization current is thus produced between two electrodes, held at constant potential in the gas. The ionization current is amplified and used to provide the chromatogram. Ionization can also be brought about by means of excited rare gases as in the Argon detector. The carrier gas used in this case is argon. Recently a new helium detector was announced which is claimed to be an ultra sensitive detector for gas analysis[†]. The detector works on the principle that helium in a high energy state is capable of ionizing all other gases except neon.

(2) INTEGRAL DETECTORS give a signal, the magnitude of which is proportional to the total quantity of a component which has passed through the detector. The oldest example of an integral detector is the titration cell, but the gas burette can also be used.

For comparison of the two types of signal see the diagrams on the opposite page →

---

* Reproduced from 'A Manual of Physical Methods in Organic Chemistry' by Sixma and Wynberg, by kind permission of John Wiley & Sons Ltd., Baffins Lane, Chichester.

† Hartmann & Dimich, *J. Gas Chrom.* **4**, 163 (1966).

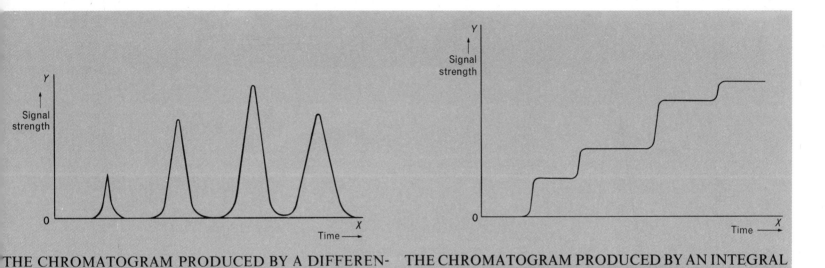

THE CHROMATOGRAM PRODUCED BY A DIFFEREN-
TIAL DETECTOR LOOKS LIKE THIS i.e. A NUMBER OF
PEAKS.

THE CHROMATOGRAM PRODUCED BY AN INTEGRAL
DETECTOR LOOKS LIKE THIS i.e. A SERIES OF STEPS.

Now turn back to ← page 52.

You think that there would be more of component A than component B. No. Have you calculated the areas correctly? Here are the peaks again.

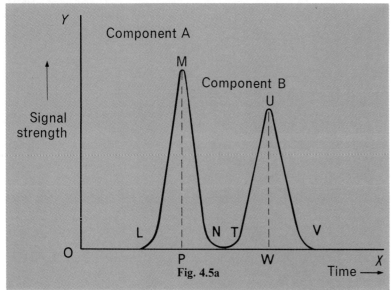

Fig. 4.5a

Area LMN is $\frac{1}{2}$(LN)(PM), where LN was given as 20 mm and PM was given as 35 mm.

Substitute these values in the formula and find area LMN. Similarly Area TUV = $\frac{1}{2}$(TV)(WU), where TV was given as 28 mm and WU was given as 25 mm.

Substitute these values in the formula and find area TUV. Are these areas equal?

What can you now say about the amounts of components A and B in the original mixture?

When you have answered these questions, return to ← page 63 and choose the correct alternative.

You think that the chromatogram would show two peaks, the retention times of which were not x min. and y min., well I'm afraid you are wrong.

Retention times are affected by two things

(i) change of column temperature
(ii) change in the volume flow rate of the carrier gas.

Has the column temperature been altered in the experiment?

Has the volume flow rate of the carrier gas been changed?

After answering these questions correctly, based on the information given, return to ← page 68 and choose the other alternative.

back ref. page 5

Quite right. In the formula, PW = 48 mm, RS = 12 mm and YZ = 20 mm, so $R = \dfrac{2(48)}{20 + 12} = \dfrac{2(48)}{32} = 3$

Now suppose our chromatogram had turned out like this!

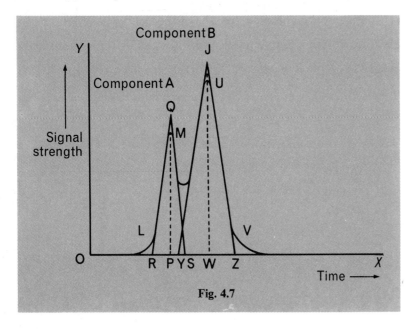

Fig. 4.7

Measurements show that PW = 16 mm, RS = 14 mm and YZ = 30 mm. Substituting in the formula for the Peak Resolution

$$R = \frac{2(PW)}{RS + YZ}$$

we see that $R = \dfrac{2(16)}{14 + 30} = \dfrac{32}{44} = 0.73$

in other words, the separation is inadequate. In general we can say that *when the Peak Resolution R is greater than* 1, *the separation is adequate*.

Question

From the formula, it will be readily seen that the best separation occur when the peak widths are narrow. In order to improve the separation either

1. the detector could be changed. ← page 57
2. a column containing a different stationary phase
   could be used. → page 77

If you have no idea what to do. ← page 62

You think that there would be more of component B than component A. No. Have you calculated the areas correctly? Here are the peaks again.

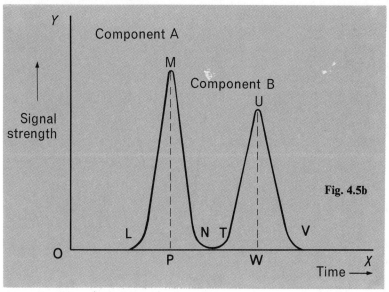

Fig. 4.5b

Area LMN is $\frac{1}{2}$(LN)(PM), where LN was given as 20 mm and PM was given as 35 mm.

Substitute these values in the formula and find area LMN. Similarly Area TUV = $\frac{1}{2}$(TV)(WU), where TV was given as 28 mm and WU was given as 25 mm.

Substitute these values in the formula and find area TUV. Are these areas equal?

What can you say about the amounts of components B and A in the original mixture?

When you have answered these questions, return to ← page 63 and choose the correct alternative.

Your answer, 1·3 is wrong. You have made an error in substituting the values given. Here is the chromatogram again.

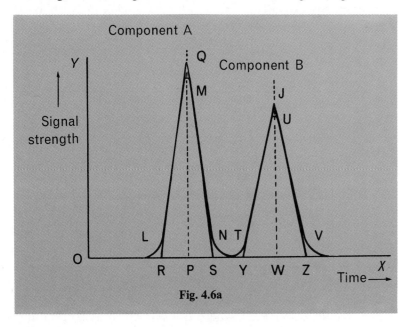

Fig. 4.6a

Now PW was given as 48 mm; RS as 12 mm; and YZ as 20 mm.

The Peak Resolution $R = \dfrac{2(\text{PW})}{\text{RS} + \text{YZ}}$

By substitution, calculate R.

Return to ← page 55 and choose the correct alternative.

You would use a column containing a different stationary phase. Correct. *The width of any peak depends only on the chromatographic process.* The only thing that is influenced by the *detector* is the *sensitivity* of the electric signal.

The relationship between the retention volume (V_{max}) and the volume of gaseous moving phase in the column (V) is expressed by the formula:

$$V_{max} = \frac{V}{p}$$

where p = fraction of the component in the gaseous moving phase.

The more soluble the component is in the stationary phase (i.e. p small) the more carrier gas will be needed to elute that component, and consequently the retention time will be long and a rounded 'peak' will appear in the chromatogram.

Conversely, the less soluble the component is in the stationary phase (i.e. p large), the less carrier gas will be needed to elute that component. Thus the retention time will be short, and a narrow, sharp peak will appear in the chromatogram.

A GOOD ANALYST BALANCES ALL THE CONDITIONS TO GET OPTIMUM RESULTS.

Turn now to → page 78.

Part Five

The Stationary Phase

BEFORE we can use any gas–liquid chromatograph, we need to
be able to select a suitable stationary phase for the separation
required. We must learn how to apply the stationary phase
uniformly to the support material, and how to pack the coated
support material into its supporting tube.

It will readily be seen that the stationary phase must be a liquid
which is chemically stable over a fairly wide range of temperature
and inert with respect to the solutes, but it must have two other
properties.

Question

Remembering that the separation of the components of a
mixture between two phases depends on their partition coefficients, would you say that the two other essential properties,
over this range of temperature were:

1. (i) low vapour pressure (i.e. <0.1 mm)
   (ii) a good solvent for the components?   → page 83

2. (i) low vapour pressure (i.e. <0.1 mm)
   (ii) a poor solvent for the components?   → page 81

3. (i) vapour pressure immaterial
   (ii) a good solvent for the components?   → page 89

4. (i) vapour pressure immaterial
   (ii) a poor solvent for the components?   → page 91

Your answer, one molecule, is wrong.

Let us look at the formula for our alcohol again. Here it is:

$$
\begin{array}{c}
R \\
| \\
O \\
\quad \diagdown \\
\qquad H
\end{array}
$$

Let's call it molecule A. Now, we can write other molecules near it, like this:

$$
\begin{array}{c}
R \\
| \\
O \longleftarrow \text{molecule A}
\end{array}
$$

Can you see how the hydrogen atoms would form hydrogen bonds with the oxygen atoms?

How many molecules are attached by hydrogen bonding to our original molecule A?

Return to → page 82 and choose the correct alternative.

Your answer, an unlimited number of molecules is wrong, I'm afraid.

Remember, the hydrogen atom forms a bridge between two electronegative atoms such as fluorine, oxygen or nitrogen.

As well as writing our carboxylic acid like this

$$R-C\underset{O}{\overset{O-H}{\lessgtr}}$$

we could also write it like this

$$\underset{H-O}{\overset{O}{\gtrless}}C-R$$

Let us look at these together. Here they are.

$$R-C\underset{O}{\overset{O-H}{\lessgtr}} \qquad \underset{H-O}{\overset{O}{\gtrless}}C-R$$

How many hydrogen atoms are available to form bridges between two electronegative atoms?

When these bridges are formed, are there any hydrogen atoms left over to form more bridges with other carboxylic acid molecules?

Answer these questions, then return to → page 109 and choose the correct alternative.

Yes, a low vapour pressure would be a desirable property but if it were also a poor solvent for the components, it would *not* be a good material for use as a stationary phase. Let us see why. The theory of partition depends on the solubility of the components of a mixture in two phases, the gaseous moving phase and the liquid stationary phase.

If the stationary phase were a poor solvent for the components then the differences in the partition coefficients would be small.

A chromatographic separation of the components of a mixture depends on making use of their differences in partition coefficients between two phases. Would a good separation be obtained if these differences were small? How could they be increased?

Return to ← page 78 and choose a better alternative.

back ref. page 97

Quite right. A carboxylic acid would be highly polar, and a hydrocarbon non polar, so an acid will not dissolve in a hydrocarbon.

Highly polar molecules will often have an attraction for each other because of the charge displacement which they each exhibit, and this tends to increase their solubility, the one in the other. This is called association, and it often takes place through a 'HYDROGEN BOND'. Let us see how this happens.

Although the hydrogen atom possesses only one electron, it is capable of forming a bond or bridge between two atoms, provided that these atoms have a strong electron attracting tendency (which is known as ELECTRONEGATIVITY). Examples of such strongly ELECTRONEGATIVE elements are fluorine, oxygen and nitrogen.

Hydrogen fluoride contains the strongly electronegative element fluorine and can be represented like this H—F. Thus the hydrogen atom in a molecule of hydrogen fluoride is able to form a hydrogen bond between two fluorine atoms so that each H—F molecule is attached by hydrogen bonds to two others like this

$$-----H-F-----H-F-----H-F$$
hydrogen bonds

and the chain could continue to an appreciable extent in the liquid form of hydrogen fluoride.

Question

Given that an alcohol can be represented by the general formula

$$\begin{array}{c} R \\ | \\ O \\ \backslash \\ H \end{array}$$ (where R can be aliphatic or aromatic)

and exhibits hydrogen bonding; would you say that each molecule of the alcohol could be attached by hydrogen bonds to:

1. one other? ← page 79
2. two others? → page 109
If you are not sure how many. → page 101

You have chosen correctly. For the theory of partition coefficients to apply, the stationary phase should be capable of dissolving the components readily, for if the solvent power was poor then the components would come off the column (i.e. be ELUTED) too quickly.

It is essential too that at the operating temperature the vapour pressure of the stationary phase should be low, to prevent that phase itself from being eluted, and to allow the components of the analytical mixture to vaporize readily.

Almost any stationary phase will give good separation of components when their vapour pressures are different, so we must consider how to separate components which have the same, or nearly the same, vapour pressures at the operating temperatures.

We can do this by using the differing affinity, (based on the nature of the chemical groups in both) which the stationary phase may have for our two components. In particular, the concept of the polarity of these groups will help.

If you have already learned about this, in relation to gas–liquid chromatography, turn to → page 115. If not, read on here.

Some ELECTRONEGATIVE elements such as fluorine, chlorine, oxygen and nitrogen have a strong charge displacement towards themselves when they combine with other atoms. When they combine with carbon to form a 'group', then that group will exhibit POLARITY, the amount of which depends on the atoms which combine together.

Question
Which of the following molecules would you expect to be POLAR?

1. benzene → page 87

2. 1 chlorobenzene → page 104

3. propane CH_3—CH_2—CH_3 → page 90

4. cyclohexane → page 96

Quite correct. You have noticed that the structure of polyethylene glycol is an ether-like structure, so our alcohol could form hydrogen bonds like this:

The result of this hydrogen bonding would be to increase the solubility of the alcohol in polyethylene glycol.

Applying this theory to gas–liquid chromatography, if we wished to separate a mixture containing an alcohol and some other polar component which could not form hydrogen bonds, then by choosing a suitable polar material like polyethylene glycol as stationary phase, hydrogen bonding would help the separation by holding back the alcohol, allowing the other component to be eluted readily.

Question

You are required to separate a mixture of ethanol (C_2H_5OH) b.p. $78 \cdot 5°C$, and acrylonitrile $(CH_2 = CH \cdot CN)$ b.p. $77 \cdot 9°C$. Which of these stationary phases would you expect to give the best chromatogram?

1. dinonyl phthalate.

→ page 8

(a semi–polar column)

2. polyethylene glycol, $(CH_2—CH_2—O—)_n$, (a polar column). → page 9

3. squalane, (a straight, long chain hydrocarbon), (a non–polar column) → page 88

If you are not sure which column to choose. → page 10

Your answer, that benzene and cyclohexane could *not* easily be separated using di-nonyl phthalate as stationary phase, is wrong! Since the boiling points of benzene (80·1°C) and cyclohexane (80·8°C) are so close, we need to make use of the greater chemical affinity that di-nonyl phthalate has for benzene, to hold back the benzene on the column and so enhance the separation due to this small difference in boiling points.

Return to → page 92 and choose a better alternative.

back ref. page 84

Remember the question asked which stationary phase would give the *best* chromatogram. Your choice, dinonyl phthalate a semi-polar phase is not correct.

It is true that the ethanol would hydrogen bond to the dinonyl phthalate and so would be held back by this association. A separation would be obtained but not the *best* separation.

Since the boiling points of acrylonitrile (77·9°C) and ethanol (78·5°C) are so close together, the more polar the stationary phase, the better the separation, and hence the better the chromatogram.

Return to ← page 84 and choose another alternative.

You think benzene is polar. You are wrong. Let us look at it again.

Polarity is conferred by the combination of carbon with a strongly electronegative element such as fluorine, chlorine, oxygen or nitrogen.

Is there any such combination in benzene?

Return to ← page 83 and choose the correct alternative.

You have chosen squalane, a non-polar column. This column would not give the best chromatogram. Let's discover why!

Since squalane is non-polar, there will be no polarity effects or holding back of any component of our mixture due to hydrogen bonding. The only criteria to be taken into account are (i) solubility and (ii) vapour pressure.

Nitriles came about half way down our table of group polarity, so they are 'semi-polar'.

Alcohols come near the top in that table and are highly polar.

Do solubility considerations help us much here?

Since the boiling points of acrylonitrile (77·9°C) and ethanol (78·5°C) are close together, what conclusion can you draw with regard to their vapour pressures at 77·9°C?

When you have answered these questions you will see why squalane will give a poor chromatogram.

Return to ← page 84 and choose a better alternative.

Yes, good solvent power is a desirable feature in a stationary phase but if it had a considerable vapour pressure then this factor would make it unsuitable, for the phase would vaporize and mix with the vapours of the components and itself be eluted.

Would this mean that the inert supporting material would be stripped of stationary phase?

What effect would this have on the value of the partition co-efficient, and thus on the chromatographic process?

Consider the answers to these questions, then return to ← page 78 and choose a better alternative.

Your answer—propane is polar. I'm afraid you are wrong.
We said that when a strongly electronegative element such as
fluorine, chlorine, oxygen, or nitrogen combines with carbon
to form a 'group', then that group will have polarity.

Let us look at propane again.

$$CH_3—CH_2—CH_3$$

Is there any polar group in propane?

Return to ← page 83 and choose the correct alternative.

I'm afraid you are wrong. Let's look at both factors.

(i) *Vapour Pressure*
If the stationary phase had a considerable vapour pressure, then it would vaporize on the column, mix with the vapours of the components of any mixture being chromatographed and be eluted.

Would this result in stripping stationary phase from the inert support material?

Would this affect partition and thus the chromatographic process?

(ii) *Solvent Power*
If the solvent power of the stationary phase was poor then the differences in the partition coefficients of the components of any mixture would be small.

Since chromatographic separation of the components of a mixture is easy when their differences in partition coefficient are large, what change in the solvent power of the stationary phase would give a good separation?

Consider your answers to the questions, and the facts given above, then return to ← page 78 and choose a better alternative.

You think that the best stationary phase would be polyethylene glycol. You are right. The ethanol would 'hydrogen bond' to the polyethylene glycol and so its solubility in that phase would increase thus holding it back on the column. This effect would enhance the separation due to their small difference in boiling points and enable the two components ethanol and acrylonitrile to be separated.

Sometimes there exists a chemical affinity between polar solvents and some non-polar solutes. This fact is particularly useful if we wish to separate a mixture of non-polar substances with close boiling points. If we choose a polar stationary phase which exhibits this affinity to one of the components, then this effect will enhance the separation due to their small difference in boiling points.

Question

Given that di-nonyl phthalate (a semi-polar material) has a greater chemical affinity for benzene (b.p. $80.1°C$) than cyclo-hexane (b.p. $80.8°C$), both non-polar materials, would you say that:

1. benzene and cyclohexane could *not* easily be separated using di-nonyl phthalate as stationary phase? ← **page 85**

2. Benzene and cyclohexane *could* easily be separated using di-nonyl phthalate as stationary phase? → **page 103**

3. Since benzene and cyclohexane are both non-polar, I would choose a suitable non-polar stationary phase for the separation? → **page 111**

Remember the question asked for the incorrect statement!
We said that polar substances dissolve in other polar substances,
and non-polar substances dissolve in other non-polar substances,
BUT a polar substance will not normally dissolve in a non-polar
substance. Here is the table of group polarity again:

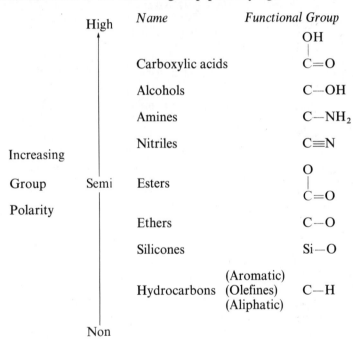

	Name	Functional Group
High		$\overset{\text{OH}}{\underset{}{|}}$
	Carboxylic acids	$C=O$
	Alcohols	$C-OH$
	Amines	$C-NH_2$
Increasing	Nitriles	$C\equiv N$
Group Semi	Esters	$\overset{O}{\underset{C=O}{|}}$
Polarity	Ethers	$C-O$
	Silicones	$Si-O$
	Hydrocarbons (Aromatic) (Olefines) (Aliphatic)	$C-H$
Non		

From this table:
Is an alcohol polar or non-polar?
Is a carboxylic acid polar or non-polar?

Answer these questions, then return to → page 97 and choose a
better alternative.

You think only two molecules would associate in this way. Yes, you are right. We could represent the association like this:

$$
\begin{array}{c}
\text{O—H} \cdots \text{O} \\
\text{R—C} \qquad \text{C—R} \\
\text{O} \cdots \text{H—O}
\end{array}
$$

Hydrogen bonds.

Carboxylic acids *can* associate in chains and large rings too, but in the gas phase the 'dimeric' form is the characteristic one.

Now as long as electronegative atoms are available they can be bridged by hydrogen bonding. Thus two dissimilar molecules could associate in this way. Let us look at an example: we can represent an ether like this $R_1\text{—O—}R_2$. If we mix the ether with an alcohol R—O—H, we could get a hydrogen bridge like this.

$$
\begin{array}{c}
\text{R} \\
| \\
\text{O} \\
| \\
\text{H} \\
\vdots \quad \text{— hydrogen bond} \\
\text{R}_1\text{—O—R}_2 \\
\vdots \\
\text{H} \\
| \\
\text{O} \\
| \\
\text{R}
\end{array}
$$

Question
Polyethylene glycol can be written like this:
—CH_2—CH_2—O—CH_2—CH_2—O—CH_2—CH_2—O—   et
Would an alcohol, R—O—H form a hydrogen bond wit polyethylene glycol?

1. Yes
2. No
If you are not sure.

← page 8
→ page 10
→ page 11

You have chosen ethanol as a non-polar molecule. Here it is again.

$$
\begin{array}{c}
\quad\;\; \text{H}\;\;\; \text{H} \\
\quad\;\; | \quad\; | \\
\text{H}-\text{C}-\text{C}-\text{OH} \\
\quad\;\; | \quad\; | \\
\quad\;\; \text{H}\;\;\; \text{H}
\end{array}
$$

It has an electron attracting group, the —OH group, attached to a carbon atom. Would this confer polarity to the molecule or not? Is the molecule symmetrical? Your answers to these questions should lead you to the conclusion that ethanol is a polar substance.

Return to → page 104 and choose another alternative.

You think that cyclohexane is polar. No. We said that when a strongly electronegative element such as fluorine, chlorine, oxygen or nitrogen combines with carbon to form a 'group', then that group will have polarity.

Here is cyclohexane.

$$\begin{array}{c}
H_2 \\
C \\
H_2C \qquad CH_2 \\
| \qquad\qquad | \\
H_2C \qquad CH_2 \\
C \\
H_2
\end{array}$$

No such 'group' exists in this molecule.

Return to ← page 83 and choose the correct alternative.

Correct. We could represent the 'pulls' like this:

$$Cl \leftarrow \overset{\overset{\displaystyle Cl}{\uparrow\uparrow}}{\underset{\underset{\displaystyle Cl}{\downarrow\downarrow}}{C}} \rightarrow Cl$$

Since the molecule is symmetrical and only carbon and chlorine atoms are involved, the **pulls** would cancel each other out, so this molecule, carbon tetrachloride, would be NON-POLAR.

It follows then, that when a molecule is *asymmetrical*, it will exhibit POLARITY. We can draw up a table of group 'polarity'.* Here it is:

		Name	Functional Group
High		Carboxylic acids	$\overset{\displaystyle OH}{\underset{\displaystyle C=O}{\vert}}$
Increasing		Alcohols	$C-OH$
		Amines	$C-NH_2$
		Nitriles	$C\equiv N$
Group	Semi	Esters	$\overset{\displaystyle O}{\underset{\displaystyle C=O}{\vert}}$
Polarity		Ethers	$C-O$
		Silicones	$Si-O$
		Hydrocarbons (Aromatic) (Olefines) (Aliphatic)	$C-H$
Non			

* This group 'polarity' is a term as used in a gas chromatographic sense and does not imply absolute polarity. It is in fact a summation of several separate effects.

The point about this list is that all polar substances will dissolve in any other polar substance. All non-polar substances will dissolve in all other non-polar substances. But a polar substance will NOT normally dissolve in a non-polar one, and vice versa.

Question

Which of the following statements is **incorrect**?

1. An alcohol will dissolve in a carboxylic acid. ← page 93
2. An ether will dissolve in another ether. → page 100
3. A carboxylic acid will dissolve in a hydrocarbon. ← page 82
4. A carboxylic acid will not dissolve in a hydrocarbon. → page 107

You have chosen the Apiezon L column. I'm sorry, but you're wrong. Let us look at the table again. Here it is:

| Component | b.p. °C | Polarity | Retention Time (mm) | |
			Apiezon L	Carbowax 20,000
Ether	35·5	low	29	17
Methyl Acetate	57	semi	30	48
Methanol	65	high	11	58

Notice that the non-polar Apiezon L column dissolves the ether (which is of low polarity) more than the methyl acetate (which is semi-polar), so that although their boiling points, i.e. 35·5°C and 57°C respectively, differ by about 22°C, their retention times are so close that the two peaks are not separated. Clearly, if Apiezon L is chosen, a difference in activity is *opposing*, rather than *assisting* a difference in vapour pressure (boiling points).

Compare the retention times of ether and methyl acetate on the Carbowax stationary phase. In this case, the difference in activity *does* assist the difference in vapour pressure (boiling points).

Return to → page 103 and choose this alternative.

Your choice, benzonitrile as a non-polar molecule, is wrong. Let us look at it together. Here it is.

$$\begin{array}{c} N \\ \parallel\parallel \\ C \\ | \\ HC{\Large\diagup}^{C}{\diagdown}CH \\ | \qquad\quad \parallel \\ HC{\diagdown}_{C}{\Large\diagup}CH \\ | \\ H \end{array}$$

The criteria for polarity are:

(i) Is there a strongly electronegative atom attached to a carbon atom, forming a 'group' which could draw electrons from the rest of the molecule towards the group?

(ii) Is the molecule symmetrical?

Your answers to these questions should lead you to the conclusion that benzonitrile is a polar substance.

Return to → page 104 and choose another alternative.

Remember the question asked for the incorrect statement. The remarks about polarity can be summarised by saying that in organic chemistry LIKE MOLECULES DISSOLVE LIKE MOLECULES.

Would you then expect one ether to dissolve in another ether?

Return to ← page 97 and choose a better alternative.

You're not sure how many molecules of an alcohol could be attached to the molecule under consideration. Let's work it out together.

We were told that an alcohol could be represented like this:

$$\begin{array}{c} R \\ | \\ O \\ \diagdown H \end{array}$$

It could also be written upside down, like this:

$$\begin{array}{c} H \\ \diagup \\ O \\ | \\ R \end{array}$$

Let's look at these together. Here they are:

$$\begin{array}{c} R \\ | \\ O \\ \diagdown \\ H \quad\quad H \quad H \\ \diagup \quad\quad\quad \diagup \\ O \quad\quad\quad O \\ | \quad\quad\quad\quad | \\ R \quad\quad\quad R \end{array}$$

Are there any hydrogen atoms available to form bridges between two oxygen atoms?

How many bridges can be formed to one molecule of an alcohol?

$$\begin{array}{c} R \\ | \\ O \\ \diagdown H \end{array}$$

Answer these questions, then return to ← page 82 and choose the correct alternative.

You have chosen chloroform as a non-polar molecule. Here it is again.

$$\begin{array}{c} H \\ | \\ Cl-C-Cl \\ | \\ Cl \end{array}$$

We could indicate the direction of movement of electrons like this

$$\begin{array}{c} H \\ | \\ \overset{\delta^-}{Cl} \leftarrow \overset{\delta^+}{C} \rightarrow \overset{\delta^-}{Cl} \\ 1. \quad \updownarrow \quad 2. \\ Cl \\ \delta^- \\ 3. \end{array}$$

as clearly there exist in the molecule three chlorine atoms which are strongly electronegative. If we number these 1, 2 and 3, the vector sum of the pulls from chlorine atoms 1 and 2 cancel each other out, leaving chlorine atom number 3 to confer polarity on the molecule.

Your choice then is wrong, for the molecule is unsymmetrical.

Return to → page 104 and choose another alternative.

You have chosen correctly. As di-nonyl phthalate has a greater chemical affinity for benzene it would hold the latter back, allowing the cyclohexane to be eluted first. As their vapour pressures at their respective boiling points are very nearly equal, this effect would enable the mixture to be separated readily.

Summarizing these effects, we can say that by careful choice of stationary phase a difference in activity can assist a difference in vapour pressure for the separation of mixtures of any solutes.

Question

An esterification mixture was extracted with ether and analyzed by gas–liquid chromatography at 160°C using a column containing Apiezon L (a non-polar substance) as stationary phase. Then the analysis was repeated using Carbowax 20,000 (a polar substance), as stationary phase. The following results were obtained.

Component	b.p. °C	Polarity*	Retention Time (mm)	
			Apiezon L	Carbowax 20,000
Ether	35·5	low	29	17
Methyl Acetate	57	semi	30	48
Methanol	65	high	11	58

Which column gave the best chromatogram?

1. Apiezon L ← page 98
2. Carbowax 20,000 → page 113
If you cannot decide which to choose → page 105

---

* Obtained from the table of Group 'Polarity', ← page 97.

Correct. The strongly electronegative chlorine will draw electrons towards itself, thus conferring a 'group' polarity on the molecule of 1 chlorobenzene.

There is an exception to this effect however. We could represent our group like this

$$\overset{\delta^+ \;\rightarrow\; \delta^-}{C\text{——}Cl}$$

where the charges of the atoms are shown. The arrow shows the direction of a small 'pull' exerted between the two atoms, due to this charge displacement. Now if these atoms were to combine with other similar atoms to form a SYMMETRICAL molecule, the various 'pulls' due to the charge displacements would balance each other. In that case despite the presence of electron attracting 'groups' in the molecule, it would have no overall polarity and would be said to be NON-POLAR.

Question

Which of the following molecules would you expect to be NON-POLAR?

1. ethanol

$$H\text{—}\underset{\underset{H}{|}}{\overset{\overset{H}{|}}{C}}\text{—}\underset{\underset{H}{|}}{\overset{\overset{H}{|}}{C}}\text{—OH}$$

← page 9

2. benzonitrile

← page 9

3. carbon tetrachloride

$$Cl\text{—}\underset{\underset{Cl}{|}}{\overset{\overset{Cl}{|}}{C}}\text{—}Cl$$

← page 9

4. chloroform

$$Cl\text{—}\underset{\underset{Cl}{|}}{\overset{\overset{H}{|}}{C}}\text{—}Cl$$

← page 10

ou can't decide which column gave the best chromatogram. et's try to decide together. Here is the table again:

| Component | b.p. °C | Polarity | Retention Time (mm) | |
			Apiezon L	Carbowax 20,000
Ether	35·5	low	29	17
Methyl Acetate	57	semi	30	48
Methanol	65	high	11	58

Remember that we said, in general LIKE DISSOLVES LIKE .e. a polar solute will be soluble in a polar solvent, and a non-olar solute will be soluble in a non-polar solvent.

ut usually a polar solute will NOT dissolve in a non-polar olvent

nd a non-polar solute will NOT dissolve in a polar solvent.

You were told that Apiezon L was non-polar, and Carbowax 0,000 was polar. Looking at the figures in the Apiezon L column ou will notice that the retention times of ether (b.p. 35·5°C) and nethyl acetate (b.p. 57°C) are so close that these peaks would not be separated. Look now at the retention time of the highly polar methanol. It is the lowest of all. These facts confirm what vas said above about polarity and solubility.

Now compare the retention times of these substances using the Carbowax stationary phase.

How has the effect of polarity on solubility affected the retention times?

You have seen that the difference in boiling points between ether and methyl acetate is almost 22°C. On which stationary phase then, does a difference in activity (i.e. solubility of substance in the stationary phase based on polarity) assist a difference in vapour pressure (boiling points) in the separation of the components?

Return to ← page 103 and choose this alternative.

You're not sure which column to choose. Let's do it together systematically.

First, let's look up the polarity of our two components, acrylonitrile and ethanol in the table of group polarity ← **page 97**. We find that acrylonitrile is semi-polar (nitriles are halfway up the table) and ethanol is highly polar (alcohols are almost at the top of the table).

Remembering what we said about polarity and solubility, the fact that both acrylonitrile and ethanol are polar materials, rules out the non-polar stationary phase, squalane.

Secondly, we must remember that at the boiling point of acrylonitrile (77·9°C) the vapour pressure of ethanol (b.p. 78·5°C) will be high, so we need to use a stationary phase which will have a high attraction for ethanol through hydrogen bonding.

Which do you think will have the greater attraction for ethanol, the semi-polar di-nonyl phthalate or the highly polar polyethylene glycol stationary phase?

Think about the answer to this question, then choose that stationary phase on ← page 84.

Remember the question asked for the incorrect statement, so you have chosen the wrong alternative.

Let us see why. Here is the table of group polarity again:

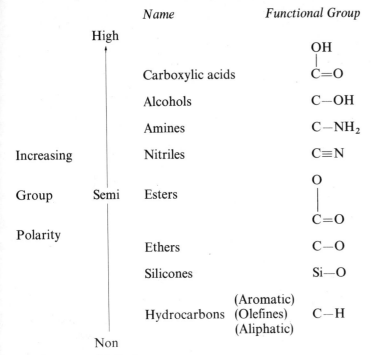

		Name	Functional Group
	High		
		Carboxylic acids	OH | C=O
		Alcohols	C—OH
		Amines	C—NH$_2$
Increasing		Nitriles	C≡N
Group	Semi	Esters	O | C=O
Polarity		Ethers	C—O
		Silicones	Si—O
		Hydrocarbons (Aromatic) (Olefines) (Aliphatic)	C—H
	Non		

Remember we said that:

Polar substances dissolve in other polar substances.

Non-polar substances dissolve in other non-polar substances.

BUT A polar substance will not *normally* dissolve in a non-polar substance.

From the table opposite:

Is a carboxylic acid polar or non-polar?

Is a hydrocarbon polar or non-polar?

Answer these questions, then return to ← page 97 and choose the correct alternative.

You think that an alcohol R—O—H could not form a hydrogen bond with polyethylene glycol. Sorry, but you're wrong.

Polyethylene glycol has ether type linkages in it holding the chain together. Here is one unit of the chain

$$-CH_2-CH_2-O-CH_2-CH_2-$$

Now, remember our alcohol R—O—H contains an electronegative oxygen atom, so clearly the hydrogen atom could form a hydrogen bond between itself and the oxygen in the glycol.

Let us look at them together, writing the alcohol like this

$$
\begin{array}{c}
R \\
| \\
O \\
| \\
H
\end{array}
$$

$$-CH_2-CH_2-O-CH_2-CH_2-$$

How many molecules of the alcohol could associate with each unit of the chain of polyethylene glycol?

Now return to ← page 94 and choose the correct alternative.

ight. Each molecule of an alcohol could be attached by hydrogen bonds to two* others, and it can be shown that in the crystals of alcohols or phenols each R—O—H unit is attached to two others in a chain like this:

with the hydrogen bonds shown by the dotted lines.

The chain could continue to an appreciable extent in the liquid form.

However, not all the compounds capable of forming hydrogen bonds do so in this unlimited way. Hydrogen bonding can link up two parts of identical molecules, provided that a hydrogen atom can form a bridge between ELECTRONEGATIVE atoms such as fluorine, oxygen and nitrogen.

---

* It is possible that an alcohol molecule could be attached by hydrogen bonds to *three* others. although the actual occurrence of this arrangement may be restricted by statistical and other factors.

Question
Given that a carboxylic acid can be represented by the general formula:

and exhibits hydrogen bonding, how many molecules would you say could associate together in this way?

1. an unlimited number
2. two
If you are not sure.

← page 80
← page 94
→ page 112

back ref. page 94

You don't know whether an alcohol, R—O—H, would form a hydrogen bond with polyethylene glycol. Let's find out. We could write our alcohol like this

R
|
O
|
H

Polyethylene glycol is made up of a large number of units. Here is one of them

—CH_2—CH_2—O—CH_2—CH_2—

Now let's look at these two together. Here they are

R
|
O
|
H

—CH_2—CH_2—O—CH_2—CH_2—

Is there a hydrogen atom available that can bridge two electronegative atoms (such as fluorine, oxygen or nitrogen) by hydrogen bonding?

From your answer to this question return to ← page 94 and choose the correct alternative.

back ref. page 92

You would choose a non-polar stationary phase for the separation. I'm afraid this wouldn't be much help.

Since at the boiling point of benzene (i.e. 80·1°C) the vapour pressure of cyclohexane (b.p. 80·8°C) would be almost 760 mm, (that of benzene at its boiling point), clearly some other effect must be used to get a reasonable separation. That effect is, that *di-nonyl phthalate has a greater chemical affinity for benzene than cyclohexane.*

This affinity effect means that using di-nonyl phthalate a good separation can be obtained between benzene and cyclohexane.

Return to ← page 92 and choose the correct alternative.

So you're not sure how many molecules of our carboxylic acid could associate together by hydrogen bonding. Let's work it out together.

We can write down our carboxylic acid like this

$$R-C\underset{O}{\overset{O-H}{\diagup}} \quad \textit{or} \text{ like this} \quad \underset{H-O}{\overset{O}{\diagup}}C-R$$

Let's look at these together. Here they are.

$$R-C\underset{O}{\overset{O-H}{\diagup}} \qquad\qquad \underset{H-O}{\overset{O}{\diagup}}C-R$$

Are there any hydrogen atoms available to form bridges between two electronegative atoms (e.g. fluorine, oxygen, nitrogen)?

How many bridges or hydrogen bonds can be formed?

When these bridges are formed, are there any hydrogen atoms left over to form more bridges with other molecules of our carboxylic acid?

Answer these questions, then return to ← page 109 and choose the correct alternative.

ight. Due to the fact that LIKE DISSOLVES LIKE, by using e Carbowax 20,000 stationary phase, a difference in activity e. polarity) is assisting a difference in vapour pressure (i.e. oiling points) for the separation of the esterification mixture.

inally, you will readily see that no permanent reactivity can be olerated between any component of the analytical mixture and e liquid phase. For example, if any component is basic, then a asic stationary phase must be used, and not an acidic one.

As a general rule we make the liquid (stationary) phase as similar in chemical class to the components of a mixture to be separated as possible. There should be no possibility of reactivity between the chemical groups of the stationary phase and those in the components of the analytical mixture.

Here is a list of a few useful stationary phases, which will do 90% of all separations.

Stationary Phase	Formula (or Functional Group)	Operating Range °C	Polarity
Polyethylene Glycol (carbowax 15 to 20,000)	$HO-CH_2-CH_2-(O-CH_2-CH_2-)_nOH$	70 to 200	polar
Polyethylene Glycol Adipate	$-CH_2-CH_2-O-CO-(CH_2)_4-CO-O-CH_2-CH_2-O-$	20 to 180	polar
Di-nonyl Phthalate	$COOC_9H_{19}$ $COOC_9H_{19}$	20 to 130	semi-polar
Silicone Fluid (MS 550)	$-O-Si-O-Si-O-$ with Me groups	20 to 200	semi-polar
Apiezon L	C_nH_{2n+2} where n is large	60 to 300	non-polar
Squalane	2,6,10,15,19,23-hexamethyl tetracosane.	0 to 150	non-polar

Part 6 starts on → page 115.

Part Six

The Preparation and Packing of Columns

THE following 26 pages are divided into five horizontal 'bands'. In each band, on consecutive right-hand pages, information is given and you will find one or more words missing, indicated by a 'dash', thus ————. You fill in the missing ———— and turn to the next page, where the correct answer(s) is given.

Do not read any further down this page, but now turn to → page 116

now turn to → page 116

12 A good ———— material should be ———— and ————. It should also have a specific area of ———— to ———— metres^2/gramme and a uniform particle size within the limits of ———— to ———— mesh. As well as ———— and ———— ————, Celite also fulfils all these conditions.

25 To ensure a ———— coating of the stationary phase on the support material, the ———— solvent is evaporated completely by carefully warming the slurry under ————, whilst shaking.

38 One simple way of ———— the tube is to hold it vertically and tap the end on the floor.

51 In winding the ———— on the ———— ————, one should aim to have the spiral as uniform as possible.

Your answer should be 'word(s)'

Now see → page 117

support, inert, porous.
2, 5, 60
100, kieselguhr, crushed firebrick

uniform, volatile
vacuum

vibrating

column, cylindrical former

Continue in this fashion, remembering to read only the top band, filling in the missing ——— in the book, or on a separate sheet of paper, until you reach page 140, where you are instructed to turn back to Section 12, which you will find in the second horizontal band on page 115.

Now turn to → page 118

13 Three materials which fulfil all the conditions for a good support material are ———, ——— ——— and ———.

26 To sum up, the two essential conditions for a ——— coating of the stationary phase on the support material are (i) that the phase should be dissolved in a sufficient volume of ——— solvent to cover the support material ——— and form a wet slurry, and (ii) the solvent must be evaporated completely by carefully warming and shaking the slurry under ———.

39 Holding the tube ——— and tapping it on the floor is a good way of ——— the tube whilst filling it with the coated support material. This can be done by hand or mechanically.

52 When glass tubing is used to hold the coated support material it must be pre-coiled before use, leaving the ends vertical and parallel as shown in Fig. 6.4. As before, one end, B, is closed with a ——— ——— ———.

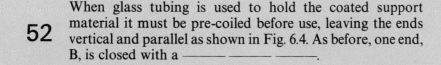
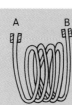

Fig. 6.4

118

word(s)

Now read Section 1 on → page 119

kieselguhr, crushed
firebrick, Celite

uniform
volatile
entirely
vacuum.

vertically, vibrating

glass wool plug.

1 The stationary phase in gas-liquid chromatography must be supported by an inert material. An _inert_ material is used to support the stationary phase.

14 Each 100 parts of support material usually contain 5 to 30 parts by weight of stationary phase absorbed on to it. The amount of stationary phase usually absorbed on 100 parts of support material is ——— to ——— parts by weight.

27 The coated support material is now packed into a suitable tube which can be made of metal, glass or plastic.
———, glass or plastic tubing can be used to hold the coated support material.

40 Other ways of ——— the tube consist of holding a Pifco Massager against the tube whilst it is held ———.

53 Slight vacuum is now applied to end B, whilst a little of the coated support material is introduced at the open end A. Columns made of glass are filled by a ——— technique.

inert

5, 30

Metal

vibrating
vertically

vacuum

2 The material used to support the stationary phase must be ——— and porous.

15 The range of ——— to ——— parts by weight of stationary phase to each ——— parts of support material is needed because the subsequent amount of solute injected depends on the amount the stationary phase can accommodate.

28 Because it is easily coiled, copper is a suitable metal tubing in which to pack the coated support material.
A suitable metal tubing in which to pack the coated support material is ———.

41 When the tube is prepared, if one of the two methods described for filling it has been chosen, the next step is to plug one end of the tube with glass wool. This end, containing a ——— of ——— will form the lower end of the tube.

54 To ensure ——— packing, the column is vibrated gently whilst filling under slight ———.

inert

5, 30, 100

copper.

plug, glass
wool

uniform, vacuum.

3 The stationary phase is supported on ——— and ——— material.

16 The amount of ——— which can be injected depends on the amount of ——— phase absorbed on to the support material. For this reason, usually ——— to ——— parts of this phase are applied to each ——— parts of support material.

29 There are other suitable metals besides ———. Aluminium and stainless steel tubing are easily ———, are quite inert, have good thermal conductivity so these metals can be used to hold the coated support material.

42 A little of the coated support material is now introduced into the upper, open end of the tube. The lower end closed with the ——— ——— ——— is now ——— by tapping on the floor or by holding the Pifco massager against the tube whilst it is held ———. The packing density of the coated support material should be about 3 g/metre length of tube.

55 The process is continued until the glass column is almost full. Then end A is closed with a ——— ——— ———, and the vacuum line is removed from end B.

inert, porous

solute, stationary
5, 30,
100

copper
coiled

glass wool plug, vibrated
vertically

glass
wool plug

4   A suitable inert material is 'kieselguhr' because it is also ———.

17   The stationary phase must be applied to the support material as uniformly as possible. The support material must be ——— coated with the stationary phase.

30   Three metals whose tubing can easily be ———, and used to hold the coated support material are ———, ——— and ——— ———.

43   The process is continued until the tube is almost full. The upper end is then closed with a ——— ——— ——— similar to that at the lower end of the tube. When full the tube should hold about ——— g/metre length of tube of the packing.

56   Glass columns and already spirally formed metal columns, then, are filled under slight ———, and are vibrated gently to ensure ——— packing.

porous

uniformly

coiled
copper, aluminium, stainless steel

glass
wool plug
3

vacuum, uniform

5 One suitable material which is both inert and ——— is called ———.

18 To apply this ——— coating, the weighed amount of stationary phase is dissolved in a sufficient volume of a volatile solvent to cover the support material entirely.

31 The dimensions of the tube may vary depending on its application and the size of the oven of the particular gas-liquid chromatography apparatus. However, the method of preparing and packing it can be seen by considering a copper tube of $\frac{3}{16}$ in. outside diameter (o.d.). Copper tubing of ——— o.d. can be used to illustrate the method of packing the coated support material into a metal tube.

44 There is one other way of filling the tube. That is to bend the tube in a loop as shown in the Fig. 6.1. With the tube held at A a little of the coated support material is introduced into one of the open ends B or C. The tube is now ——— by tapping it on the floor at D whilst the tube is held ———. As in the 'straight' methods of filling the tube, the amount of packing taken is about ——— ———/——— length of tube.

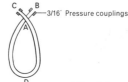

Fig. 6.1

57 No matter what material is used to hold the coated support material, the column when wound, must be 'conditioned' before use. Before any column can be used it must be '———'.

porous, kieselguhr

uniform

$\frac{3}{16}$ in.

vibrated
vertically
3g/metre

conditioned

6 The first support material we have learned about is called ———.

19 By dissolving the stationary phase in a sufficient volume of a ——— solvent to cover the support material ———, a uniform coating may be obtained.

32 In this case, for the value of H of the column (← **page 66**) to be about 1 mm a column length of about 1 to 2 metres is usually used. The normal length of the column is ——— to ——— metres.

45 As before, the process is repeated until one limb is full. This limb is then closed with a ——— ——— ———. The other limb is filled and then closed in the same way.

58 The reason why a column must be ——— is to remove any excess solvent or volatile materials from the stationary phase before use.

kieselguhr

volatile,
entirely

1, 2

glass
wool plug

conditioned

7 There are other possible support materials besides ———. Crushed firebrick is both ——— and ——— so it can be used as a support material.

20 Some solvents for the stationary phase are pentane, di-ethyl ether or acetone. These are sufficiently ———, to enable a ——— coating of stationary phase to be applied to the support material.

33 The dimensions of the copper tubing used to hold the coated support material are ——— o.d. and a length of ——— to ——— metres.

46 There are ——— ways of packing the coated support material into the ——— o.d. copper tubing once it is fitted with the ——— ——— couplings. In each the tube is held ——— whilst being filled and to ensure ——— packing the tube is ——— by tapping it on the floor or by using a Pifco Massager. The ends of the tube when full, are closed with ——— ——— ———. When full the packing density should be ——— ———/——— length of tube.

59 When we say a column has been ———, we mean that any excess ——— or ——— materials have been removed from the stationary phase.

kieselguhr, inert
porous

volatile, uniform

$\frac{3}{16}$ in.
1, 2,

three, $\frac{3}{16}$ in.
$\frac{3}{16}$ in. pressure, vertically
uniform, vibrated
glass wool plugs
3 g metre

conditioned, solvent, volatile

8 The ———— material must also have a specific area of 2 to 5 metres^2/gramme.

21 ————, ———-———— ———— or ———— are suitable solvents for the stationary phase, which must be of sufficient volume to cover the support material ————, to enable the stationary phase to be ———— coated on it.

34 As the column works under pressure and needs to be connected to the rest of the system, $\frac{3}{16}$ in. pressure couplings* are fitted to each end of the copper tube. To enable the column to be connected to the rest of the chromatographic system under pressure $\frac{3}{16}$ in. ———— couplings are used.

47 The tube when packed with the coated support material is called a column. The ———— needs to be wound into a spiral form to allow it to fit conveniently into the oven of our gas–liquid chromatography apparatus.

60 To carry out this conditioning process, the column is connected to the chromatographic apparatus, and carrier gas is passed through it. The first step then in the conditioning process is to pass———— gas through it. In order not to contaminate the detector, the exit end of the column should not be connected to it during this process.

* See Appendix B page → **292**

support

pentane, di-ethyl ether, acetone
entirely
uniformly

pressure

column

carrier

9 A material with a specific area of ——— to ——— metres^2/gramme may be used as a ——— material.

22 When the stationary phase, support material and ——— solvent are mixed together and shaken, a wet slurry is formed.

35 Features of a suitable tube to hold the coated support material are (i) an o.d. of———, (ii) a length of ——— to ——— ———, (iii) fitted with ——— ——— couplings at each end.

48 To wind the ——— into a spiral form, a cylindrical former is used. Fig. 6.2 shows what this looks like.

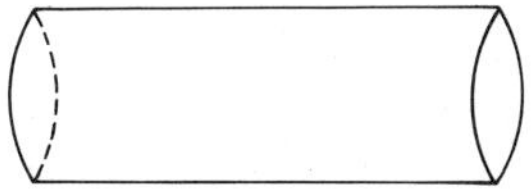

Fig. 6.2

61 As the ——— gas is passing, the column is heated in the g.l.c. oven to a temperature about 10°C below the maximum temperature recommended for that particular stationary phase.

2, 5, support

volatile

$\frac{3}{16}$ in.
1, 2 metres, $\frac{3}{16}$ in. pressure

column

carrier

10 Not only should the support material have a specific area of ——— ——— ——— ——— ——— ——— but it should have a uniform particle size within the limits of 60 to 100 mesh.

23 The wet slurry consists of stationary phase, ——— material and ——— solvent. The solvent is then evaporated by carefully warming the slurry, with shaking, under vacuum.

36 When the tube is so fitted, all that remains is to pack the coated support material uniformly into the tube. ——— packing is advisable when filling the copper tube with the coated support material.

49 The diameter of the cylindrical ——— should be such as to allow the column, when wound, to fit conveniently into the oven of our gas–liquid chromatography apparatus, the diameter of the ——— ——— should not be smaller than 10 times the diameter of the column tubing.

62 The column, which is being heated to a temperature about ——— °C lower than the maximum temperature recommended for that particular stationary phase, whilst ——— gas is passing, is kept under these conditions for about twelve hours. It is essential during this process not to connect the ——— to the ——— in order not to contaminate the latter.

2 to 5 metres^2/
gramme

support, volatile

Uniform

former
cylindrical former

10
carrier
column, detector

11 Uniform particle size between the limits of ——— to ——¹⁰⁰ mesh is a desirable feature of a support material. The closer the mesh size within these limits, say 90 to 100 mesh, the greater the efficiency of the column.

24 As the solvent evaporates under ———, the shaking movement ensures a ——— coating of the stationary phase on the support material.

37 To achieve ——— packing, the tube must be vibrated in some way during the filling operation.

50 Fig. 6. 3 shows the ——— wound on the ——— ———. The diameter of the latter should not be smaller than ——— times the diameter of the ——— tubing.

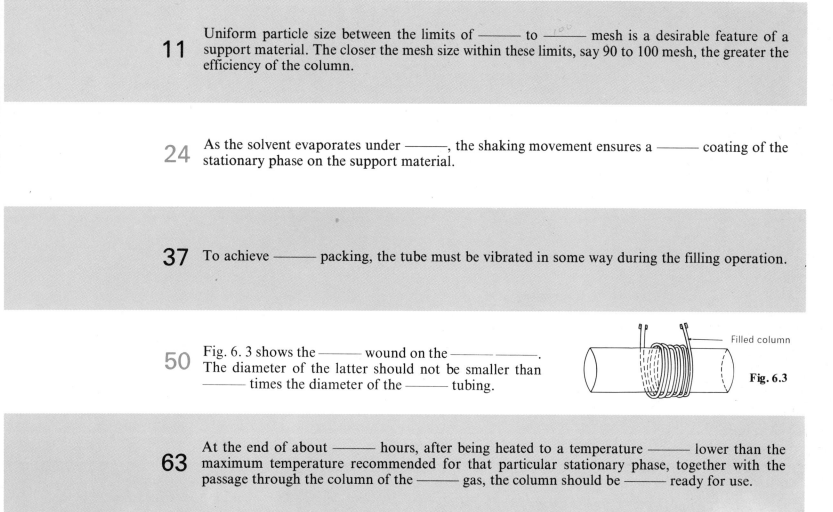

Filled column

Fig. 6.3

63 At the end of about ——— hours, after being heated to a temperature ——— lower than the maximum temperature recommended for that particular stationary phase, together with the passage through the column of the ——— gas, the column should be ——— ready for use.

60, 100   **Turn back to section 12 on ← page 115.**

vacuum, uniform   Turn back to section 25 on ← page 115.

uniform   **Turn back to section 38 on ← page 115.**

column, cylindrical former
10, column   Turn back to section 51 on ← page 115.

twelve, 10°C

carrier, conditioned   **Turn now to → page 141**

Having read through the sequence on 'The Preparation and Packing of Columns' write down the answers to the following questions in your notebook:

Questions

1. Give any three of the four desirable properties of a supporting material for the stationary phase.

2. Which property increases the efficiency of the column?

3. How many parts by weight of the stationary phase are usually applied to 100 parts by weight of the support material?

4. How would you ensure that the supporting material was uniformly coated with the stationary phase?

5. What size of copper tubing is normally used to contain the column packing?

6. Why is it advisable to vibrate the tube during the filling operation?

7. Give the recommended packing density for any length of column.

8. When the column is completed, name and describe what remains to be done before it can be used for analysis.

Now turn to → page 283

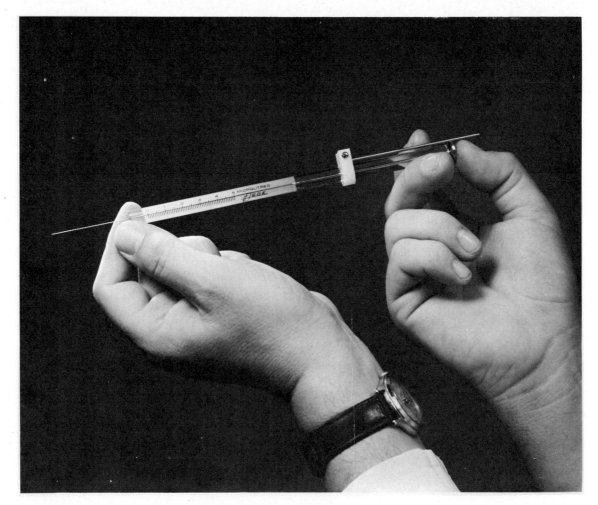

The correct way to hold a syringe

Part Seven

Sample Injection and Syringe Technique

Work through these sections in the same way as those in Part Six.

1 Having seen how to prepare a column, we need to learn how to inject a sample of our solute mixture on to it. ———— injection is done with a precision 10 microlitre syringe.

6 This liquid in the ———— of the syringe is then emptied by tapping the top of the plunger, whilst the ———— is held below the surface of the solute mixture. This process is repeated until all the air bubbles have been removed from the syringe.

11 We can now say that there are two precautions to be taken before making an injection. The first is that all ———— ———— must be removed from the barrel. The second is that the ———— of the ———— must be wiped clean with absorbent paper after sampling the solute mixture. These precautions ensure that only the ———— ———— of solute taken into the ———— of the syringe, finds its way on to the column.

16 The sequence of operations can be summarised as follows. (i) A little of the solute mixture is drawn into the ———— of the syringe. The ———— ———— are then removed from the ————. (ii) A sample of a ———— ———— of the solute mixture is taken. (iii) The ———— of the ———— is wiped clean with absorbent paper, then the ———— is inserted all the way into the ———— ———— of the g.l.c. apparatus. (iv) The ———— is depressed rapidly. (v) The ———— point is marked on the recorder chart and at the same time the ———— is withdrawn rapidly from the ———— ————.

21 The first of these is, to avoid bending the needle or the plunger at all times. ———— the needle or the plunger would obviously make accurate injections difficult, if not impossible.

Sample

barrel
needle

air bubbles, outside
needle
few microlitres, barrel

barrel, air bubbles, barrel
few microlitres, outside, needle
needle, rubber septum
plunger, injection
needle, rubber septum.

Bending

2   A precision 10 microlitre ———— is used to inject a ———— of our solute mixture on to our column.

7   For accurate work, all the ———— bubbles must be removed from the ———— of the syringe. This is done by taking some liquid in the syringe, then emptying it by ———— the top of the plunger whilst the ———— is held below the surface of the solute mixture.

12   After cleaning the ———— of the ———— with absorbent paper, the ———— is now inserted all the way into the rubber septum of the injection port of the g.l.c. apparatus, and emptied by rapidly pressing the plunger.

17   After the injection the needle and barrel of the syringe are cleaned with a solvent such as di-ethyl ether. A useful solvent for cleaning the syringe internally is ———— ———— ————.

22   As well as not ———— the needle or the plunger, never touch any part of the plunger except the top. Dirt and skin grease from your fingers would rub off on to the plunger. This ———— and ———— grease would quickly cause the plunger to seize up in the syringe barrel.

syringe, sample

air, barrel
tapping
needle

outside, needle, needle

di-ethyl ether

bending
dirt, skin

3 Fig. 7.1 (not to scale) shows a precision 10 microlitre ————.

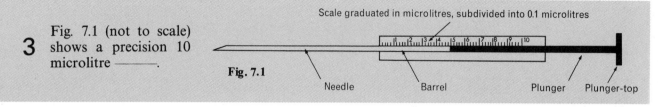

Scale graduated in microlitres, subdivided into 0.1 microlitres

Fig. 7.1

Needle   Barrel   Plunger   Plunger-top

8 Once the ———— has been removed from the ———— of the syringe, the required amount, usually a few microlitres (written µl) or less can be drawn into the syringe————.

13 Sample injection with a ———— ———— ———— ———— is carried out by inserting the ———— all the way into the rubber ———— of the injection port, and emptying it rapidly by pressing the ————.

18 The excess ————-———— ———— must be removed before the next sample of solute mixture is taken. To do this the ———— of the syringe is inserted into a rubber serum cap attached to a line under vacuum. The vacuum removes the excess solvent.

23 The final precaution is to store the syringe in its case, preferably away from contamination by dust or chemicals. Storing the syringe in its ———— avoids further contamination by ———— or ————.

syringe.

air, barrel
barrel

precision 10 microlitre syringe, needle,
septum, plunger

di-ethyl ether
needle

case, dust,
chemicals.

4   To use the ——— 10 ——— ——— the needle is inserted into the solute mixture and some of the liquid is drawn into the barrel.

9   The amount usually injected is a ——— microlitres. This is drawn into the syringe barrel once all the ——— has been removed from it.

14   Name the parts of the precision 10 microlitre syringe shown in Fig. 7.2.

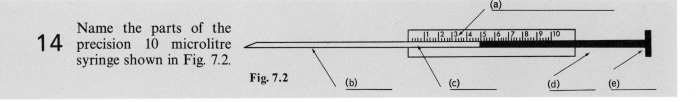

Fig. 7.2

19   Before re-using the syringe it must be cleaned using a solvent such as ———-——— ———. The precaution must be taken to remove any excess solvent by inserting the ——— into a rubber serum cap, which is attached to a line under ———.

24   Summarising these precautions we can say that at all times avoid ——— the needle or plunger. Do not touch any part of the plunger, except the top to avoid depositing ——— and ——— ——— on the plunger. Finally, store the syringe in its ——— to avoid further contamination by ——— or ———.

precision, microlitre syringe

few
air

(a) scale graduated in 0·1 μl, (b) needle
(c) barrel (d) plunger (e) plunger top.

di-ethyl ether
needle
vacuum.

bending
dirt, skin grease
case, dust
chemicals.

5 The first step in taking a ——— of the solute mixture, using the ——— ——— ——— ——— is to insert the needle into the mixture and draw some liquid into the barrel.

10 Once the sample of the solute mixture (usually a ——— ———) has been taken, the needle may be withdrawn from the mixture. The outside of it is then wiped clean with absorbent paper. This precaution ensures that only the sample of the solute mixture *inside* the syringe, finds its way on to the column.

15 After the needle has been emptied into the injection port by rapidly pressing the ———, the injection point is marked on the recorder chart and at the same time the needle is withdrawn from the rubber ———. The whole sequence of injection, marking the injection point on the recorder chart, and the withdrawal of the needle must be done as quickly as possible.

20 As well as removing excess solvent under ——— three other precautions are necessary when using a ——— ——— ——— ———.

25 If your injection has been successful a chromatogram will now be obtained. The obtaining of a ——— depends on a successful ———.

sample, precision 10 microlitre syringe **Turn back to section 6 on ← page 143.**

few microlitres Turn back to section 11 on ← page 143.

plunger

septum **Turn back to section 16 on ← page 143.**

vacuum
precision 10 microlitre syringe Turn back to section 21 on ← page 143.

chromatogram, injection **Turn now to → page 153.**

Write down the answers to the questions below in your notebook. They are based on the sequence 'Sample injection and syringe technique' which you have just read.

Questions
1. What is the name of the syringe which is normally used in g.l.c. analysis?

2. What is the usual amount of the solute mixture injected?

3. Which **two** precautions must be taken before making an injection?

4. Describe briefly how to inject the sample into the injection port, and when to mark the injection point.

5. Which useful solvent is used for cleaning the syringe internally after an injection?

6. How is the excess solvent removed?

7. Give **three** other precautions to be taken when handling and storing the syringe.

8. What is the proof of a successful injection?

Turn now to → page 284.

Part Eight
Qualitative Analysis

HAVING discussed the elementary theory of gas–liquid chromato-graphy, learned how to prepare and pack any column and inject our sample on to it we need to see how the various parts of the apparatus work together to give a chromatogram. Here is a simple 'block' diagram of the apparatus.

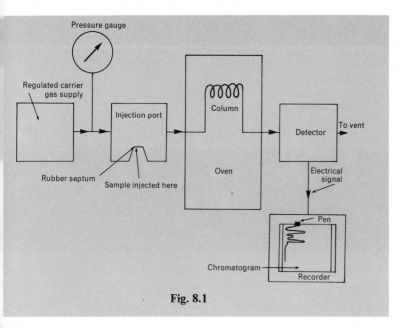

Fig. 8.1

Once the appropriate column has been selected for the analysis, *the sample is injected* into the injection port where it mixes with the carrier gas and is carried over into the column located in the oven. (On some machines the sample can be injected directly on to the column.)

On the column, *chromatographic separation* takes place, and *the detector responds* to each component as it is eluted by giving an electrical signal. The recorder displays this signal as a *visual trace*.

Electricity is needed to provide the power for various parts of the equipment.

Now turn to → page 166.

The *carrier gas supply* is regulated to a pressure of approximately 5 to 10 psi, depending on the type of g.l.c. machine being used.

back ref. page 284

Before we can proceed with the analysis two further things need to be done. First we must assess the nature of the components of the mixture. For instance the physical properties such as boiling point, polarity and/or affinity considerations and the chemical functional groups present all must be taken into account, and from these facts a suitable stationary phase is chosen.

The boiling points of the components will give us a guide as to their volatility, and, as the value of the partition coefficient depends on temperature, will help us to determine how much stationary phase to apply to the support material.

Let us see how to do this. If we were to inject a sample of air into our g.l.c. apparatus it would simply mix with the carrier gas and pass through the column without dissolving at all in the stationary phase. On elution it would give a peak, which we call the **AIR PEAK**. Let its retention time be x minutes.

Question

If we now inject on to the column a sample of our mixture, would you expect the retention times of the components to be:

1. greater than x min ?   → page 172
2. less than x min ?   → page 167

If you have no idea what the retention times would be,   → page 160

Your answer A = 2 minutes; B = 3 minutes; C = $4\frac{1}{2}$ minutes; is wrong.

We can write, for each component:

Retention time relative to the Air peak
            = Retention time − Retention time of the air peak

The retention time of the air peak was 2 minutes.

The retention time for component A was 4 minutes; for B, 6 minutes; for C, 9 minutes.

Substitute these values in the equation to obtain the retention time of each component relative to the air peak.

Now turn again to → page 172 and choose the correct alternative.

Your answer, 250, is wrong

It shows that you have calculated the value of the fraction $\dfrac{y}{(x + y)}$

to be $\frac{1}{2}$, but have forgotten that in the formula for N, the effective number of theoretical plates, this fraction is squared.

Recalculate N from the formula

$$N = n \left(\frac{y}{x + y}\right)^2$$

Then return to → page 184 and choose that value from those listed there.

No, if the amount of stationary phase is decreased, the amount of the component which it can dissolve will decrease. The result of this will be that the component will pass through the column even more quickly than before.

Return to → page 173 and choose the other alternative.

You have no idea what the retention times would be. Let's look at the facts again.

We know that if we inject a sample of air into our g.l.c. apparatus, it would pass through the column and into the detector without dissolving at all in the liquid stationary phase, whereas a sample of our mixture would dissolve in the stationary phase.

Would you expect the components of our mixture to pass through the column more quickly or less quickly than air?

Think about this, then return to ← page 156 and choose the correct alternative.

back ref. page 172

Your answer was neither of those given. Well, let's work it out together. The chromatogram could look like Fig. 8.3A.

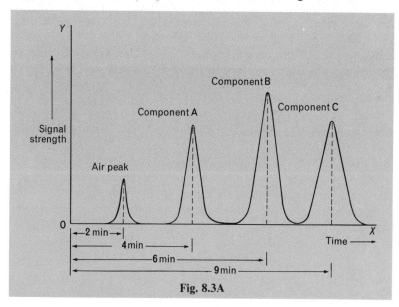

Fig. 8.3A

We can write for each component,

Retention time relative to the Air peak
= Retention time − Retention time of the Air peak

The retention times of the components were given as A = 4 minutes; B = 6 minutes; C = 9 minutes. The retention time of the air peak was 2 minutes.

Substitute these values in the equation to obtain the retention time of each component relative to the air peak, then return to → page 172 and choose the correct alternative.

Your answer was none of the alternatives listed. Let's go through the calculation together.

We were given that x, the retention time of the air peak was 2 minutes, and that $x + y$, the retention time of the single substance was 4 minutes. This means that the retention time of the single substance, relative to the air peak (i.e. y) is $4 - 2 = 2$ minutes.

Now substitute these values in the factor $\left(\dfrac{y}{x + y}\right)^2$ and calculate

its value.

We were given that the number of theoretical plates $n = 500$, so now substitute the values found and the facts given in the formula

$$N = n \left(\frac{y}{x + y}\right)^2$$

to find N, the *effective* number of theoretical plates.

Now return to → page 184 and choose the corresponding alternative.

Your choice, a medium length column, would not produce the chromatogram as quickly as possible.

Both nonane (b.p. 150·8°C) and *cis*-decalin (b.p. 195°C) have quite high boiling points, so they are of LOW volatility. Even at 75°C, the temperature of the analysis, their vapour pressures would be quite low. What effect would this have on their retention time?

What length of column then, from those listed, would give a chromatogram as quickly as possible?

Consider the answers to these questions then return to → page 174 and choose this alternative.

You would choose neither of the alternatives given.

Perhaps you think that a lower temperature than either the boiling point of the highest boiling component *or* the safe maximum operating temperature of the stationary phase would be the *first* choice of oven temperature.

If you think like this, then it could just happen that the vapour pressure/temperature curves could cross as shown in Fig. 8.5

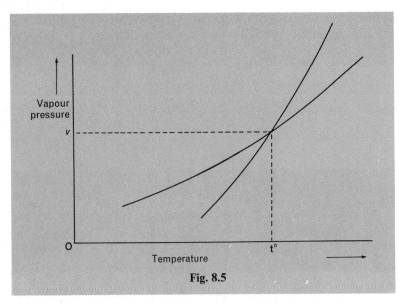

Fig. 8.5

so that at your chosen temperature of $t°$ their vapour pressures are identical. This would not make the chromatographic separation easier.

You are right in rejecting as oven temperature, the boiling point

of the highest boiling component of the mixture, for this temperature could well be above the safe maximum operating temperature of some stationary phases.

I hope then that you have come to the conclusion that the only logical *first* choice of oven temperature would be the safe maximum operating temperature of the particular stationary phase in use.

Return to → page 170 and choose that alternative.

Correct, a short column would give you a chromatogram as quickly as possible, for even at 75°C, the oven temperature for the analysis, both nonane and *cis*-decalin would have a low vapour pressure, so their vapours would pass more quickly through a short column than a long one.

Let's look again at our example. We wished to analyse by gas–liquid chromatography, a mixture of nonane (b.p. 150·8°C) and *cis*-decalin (b.p. 195°C) and the oven temperature quoted was 75°C, the stationary phase squalane. From our table of stationary phases (← page 113) we see that the maximum operating temperature of our stationary phase is 150°C. From vapour pressure citations we can make up a table as follows:

Compound	Vapour pressures at	
	75°C	150°C
Nonane	80 mm	760 mm
cis-Decalin	25 mm	400 mm

(the vapour pressures are approximate figures)

Question

If in the gas–liquid chromatography analysis of our mixture of nonane (b.p. 150·8°C) and *cis*-decalin (b.p. 195°C) we had used an oven temperature of 150°C (i.e. the maximum for squalane, the stationary phase), would the retention times of our two components have

1. increased   → page 188
2. decreased   → page 170
3. stayed the same   → page 177

provided the volume flow rate of the carrier gas was kept constant?

Write down the answers to the questions below in your notebook without reference to page 155. They are set to test your understanding of the function of the various pieces of equipment that together make up a Gas–Liquid Chromatograph.

Questions

1. Name the various parts of the apparatus shown in the block diagram (Fig. 8.2).
2. What is the regulated pressure of the carrier gas supply?
3. Assuming the appropriate column has been fitted, describe what happens after the sample has been injected, concluding your account with the production of a visual trace.
4. Which two services are essential before the equipment can be used for qualitative analysis?

Now turn to page → 284.

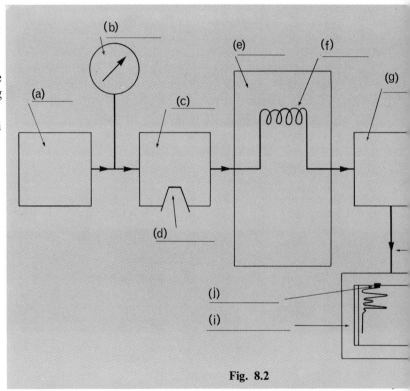

Fig. 8.2

Remember we said that our sample of air would simply mix with the carrier gas and pass through the column without dissolving at all in the stationary phase.

Would you expect the components of our mixture to dissolve in the stationary phase as they passed through the column?

Would they, then, pass through the column more quickly or less quickly than air?

Return to ← page 156 and choose the correct alternative.

You're undecided about the right amount of stationary phase to be used. Let's work it out together.

We need two pieces of information:

(i) the value of x, the retention time of the air peak. This was given as 1 minute.

(ii) the value of y_1 and y_2, the retention times of each component relative to that of the air peak.

We were given the retention times of our two components, i.e. $(x + y_1)$ and $(x + y_2)$. They were 3 and 3·5 minutes respectively.

Calculate y_1 and y_2 from this information.

Now we said that if

$y > 7x$ less stationary phase should be used (i.e. down to about 5 $w/w\%$)
$y \rightarrow x$ more stationary phase should be used (i.e. up to about 30 $w/w\%$)
y lies between x and $7x$ the amount of stationary phase was probably adequate.

Where do your calculated values of y lie?

Go back to → page 186 and select the correct alternative.

You think that the chromatogram proves that components B and C are identical. Let us look carefully at the chromatogram (Fig. 8.7A).

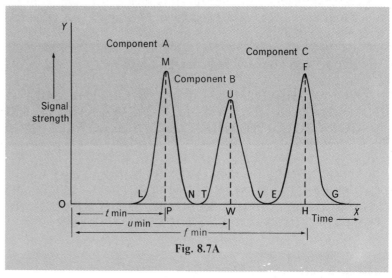

Fig. 8.7A

If components B and C *had* been the same, the effect would have been to increase the amount of component B in the detector. What effect would this have had on the peak height of component B? Has the peak height of component B changed from that height shown on → page 176?

What does this fact alone, indicate?

Assuming that $t \neq u \neq f$, what do these retention times prove?

Return to → page 190 and choose a better alternative based on the evidence above.

Yes, by raising the temperature of the oven to the maximum operating temperature of the stationary phase, we have increased the vapour pressure of our components. Since all other parameters have been kept constant, this means that the retention times will have decreased, over these obtained at an operating temperature of 75°C.

So then, we can now say, in general:

(i) WHEN THE BOILING POINTS OF THE SOLUTE COMPONENTS ARE HIGH (I.E. LOW VOLATILITY) USE A SHORT COLUMN: WHEN THEY ARE LOW (I.E. HIGH VOLATILITY) USE A LONG COLUMN.*

(ii) AN INCREASE IN THE OVEN TEMPERATURE UP TO THE STATED SAFE MAXIMUM FOR THE STATIONARY PHASE IN USE WILL RESULT IN DECREASED RETENTION TIMES, AND A QUICKER ANALYSIS.†

---

* As an alternative it *would* be possible to use a column containing less stationary phase for the separation of compounds of low volatility (i.e. high boiling materials), but this would probably result in poor separation of the peaks. But see D. H. Frederick, *et al., Anal. Chem.* **34**, 1521 (1962)
† Provided that any component of the mixture does not decompose at that maximum temperature.

Question

Suppose for any analysis, the quantity of stationary phase to use has been decided, the column prepared and conditioned, and connected to the rest of the gas–liquid chromatograph apparatus, would you say that the FIRST choice of oven temperature, for our first attempt to produce a chromatogram would be:

1. the boiling point of the highest boiling component of the mixture? → page 17⁸

2. the safe maximum operating temperature of the stationary phase? → page 17⁶

3. neither of these? ← page 16⁴

Your answer, 225, is wrong.

By arriving at this answer you have shown that your mistake was made in the calculation of the relative retention time y. The retention time of the air peak (x) was 2 minutes. The retention time of the single substance ($x + y$) was 4 minutes. This gives the value of y to be 2 minutes.

Substitute these values in the equation $N = n\left(\dfrac{y}{x + y}\right)^2$.

Now return to → page 184 and choose the corresponding value for N, the effective number of theoretical plates.

Your answer, greater than x minutes, is correct, for as the two components pass over the stationary phase they will dissolve, each to an extent depending on its own partition coefficient for the system. This means that the retention times are bound to be longer than that of air which does not dissolve in the stationary phase.

The chromatogram could look like Fig. 8.3.

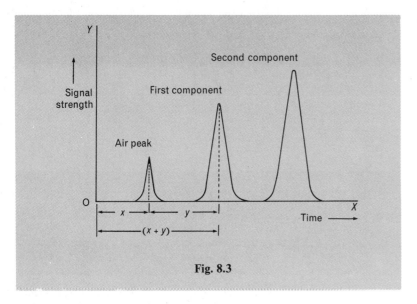

Fig. 8.3

Let us say the retention time of our first component was $(x + y)$ minutes. You will see from the chromatogram that y minutes is the retention time of the first component RELATIVE to the retention time of the Air peak.

Question

A mixture of three components, A, B and C was analysed by gas–liquid chromatography and the retention times of the peak were, for component A, 4 min, component B, 6 min and component C, 9 min. The retention time of the air peak was 2 min A calculation of the retention time of each component relative to the air peak shows it to be:

1. A = 2 min; B = 3 min; C = $4\frac{1}{2}$ min.
2. A = 2 min; B = 4 min; C = 7 min.
3. Neither of these.

← page 157
→ page 184
← page 161

quite right. From the information given $x = 2$ minutes and $(x + y) = 4$ minutes so $\left(\dfrac{y}{x + y}\right)^2 = \left(\dfrac{2}{4}\right)^2 = \dfrac{1}{4}$

substituting in the formula, $N = 500 \left(\dfrac{1}{4}\right) = 125$.

We can write a table relating y, $x + y$, and $\left(\dfrac{y}{x + y}\right)^2$

y	$x + y$	$\left(\dfrac{y}{x + y}\right)^2$
$y = 0\cdot1x$	$1\cdot1x$	$0\cdot008$
$y = 0\cdot2x$	$1\cdot2x$	$0\cdot027$
$y = 0\cdot3x$	$1\cdot3x$	$0\cdot053$
$y = x$	$2x$	$0\cdot250$
$y = 2x$	$3x$	$0\cdot445$
$y = 3x$	$4x$	$0\cdot562$
$y = 10x$	$11x$	$0\cdot826$
$y = 20x$	$21x$	$0\cdot906$
etc.	etc.	etc.

since $N = n \left(\dfrac{y}{x + y}\right)^2$ for a given column of n theoretical plates, N can be calculated readily for given values of x, and $x + y$. It will readily be seen from this table that as $x + y$, the retention time of a component, approaches x, the retention time of the air peak, the value of N decreases.

Question

For one particular component, the effective plate number will more nearly approach the theoretical plate number as the retention time of that component increases. Assuming constant temperature and volume flow rate of the carrier gas, one way of increasing the retention time of our component would be to:

1. increase the amount of stationary phase on the support material → page 186

or 2. decrease the amount of stationary phase on the support material ← page 159

back ref. page 18

Right. Your calculations have shown that $\frac{y_1}{x}$ is $\frac{(3-1)}{1} = 2$ and

$\frac{y_2}{x}$ is $\frac{(3 \cdot 5 - 1)}{1} = 2 \cdot 5$. Since these figures lie between 1 and 7 the amount of stationary phase used is probably adequate at 10 $w/w\%$.

The next factor to consider is connected with the volatility of a solute. We have seen earlier that the vapour pressure of a liquid at its boiling point is 760 mm. Consequently if a liquid has a HIGH boiling point, we say that it is of LOW VOLATILITY e.g. Dodecane b.p. 216°C. Conversely, if it has a LOW boiling point it has HIGH VOLATILITY, e.g. ether b.p. 35°C.

Since the chromatographic process depends on the partition of the solute vapour between the gas and liquid phases, then the vapour pressure of that solute at a given temperature, would affect the speed at which the vapour would pass through a column at that temperature. If the solute had HIGH VOLATILITY, its vapour pressure would be high at elevated temperature, so the vapour would tend to pass QUICKLY through the column. Conversely, if the solute had LOW VOLATILITY, the vapour would tend to pass SLOWLY through the column.

Question

If you were given a mixture of nonane (b.p. 150·8°C) and *cis*-decalin (b.p. 195°C) to analyse, using squalane (non-polar) as stationary phase, would you use a

    1. long (say 2 metres)   → page 182
or 2. medium (say 1·5 metres)   ← page 163
or 3. short (say 1 metre)   ← page 165

length of column to obtain a chromatogram as quickly as possible, assuming that the analysis was carried out at an over temperature of 75°C, with a constant volume flow rate of carrier gas?

You think that the chromatogram shows that components **B** and **C** are not identical. I'm afraid you're wrong. Here is the chromatogram again. (Fig. 8.8A).

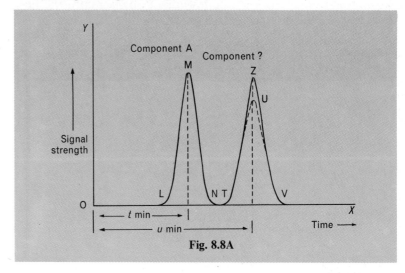

Fig. 8.8A

Is there a peak at retention time u minutes (that of component B)?

Has any peak in the chromatogram increased in peak height whilst peak width has remained constant, showing that more of that component has passed through the detector?

Are there more than two peaks in the chromatogram?

Use the evidence of the chromatogram and the answers to these questions to choose the correct alternative on → page 198.

Your choice, the safe maximum operating temperature of the stationary phase, is quite correct. It may not be the *best* temperature for the separation, but it is the *first* choice, for it will give a chromatogram. It can then be lowered, stepwise, until the best separation of the components is made.

Now look at the chromatogram shown in Fig. 8.6.

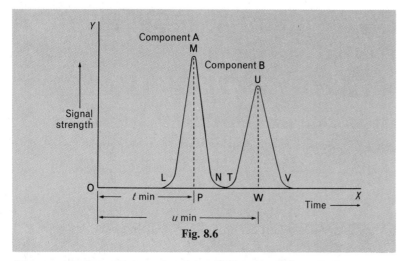

Fig. 8.6

You remember that, as long as all the conditions (temperature, carrier gas pressure, etc.) remain constant, the retention volume and consequently the retention time will always be constant. Let us say these are t minutes for component A and u minutes for component B.

The second point to remember is that the *amount* of each component present is given by the area under its peak. For component A this is area LMN, for B, area TUV.

Question

As long as the peaks in a chromatogram are of similar SHAPE would you say that we can *approximate* the area of each peak most conveniently to:

1. the peak heights i.e. PM and WU
2. the peak width i.e. LN and TV.

→ page 190
→ page 183

If you do not think it wise to try to approximate the area

→ page 187

Wrong. For the retention times to stay the same, there would have to be no change in the vapour pressure of the components. Provided that the volume flow rate of the carrier gas were kept constant, this could only happen if the temperature were kept constant.

Look at the table again:

Compound	Vapour pressures at	
	75°C	150°C
Nonane	80 mm	760 mm
cis-Decalin	25 mm	400 mm

The vapour pressure of each component has increased on doubling the temperature. As the vapour pressure increases, would you expect the chromatographic process to speed up or slow down?

What effect would such a change have on the retention time of each component?

When you have thought about the answers to these questions, turn again to ← page 165 and choose a better alternative.

At first sight it might seem that the first choice of oven temperature would be the boiling point of the highest boiling component of the mixture to be analysed, but this is not so.

Due consideration will show that this is an arbitrary method to choose as that boiling point could be relatively low or quite high. Some components even boil at temperatures higher than the safe maximum operating temperatures of some stationary phases. In such cases what does one do then?

Suppose one then elects a lower temperature than the boiling point of the highest boiling component. It could just happen that the vapour pressure/temperature curves could cross as in Fig. 8.4, so that at $t°$ their vapour pressures are identical.

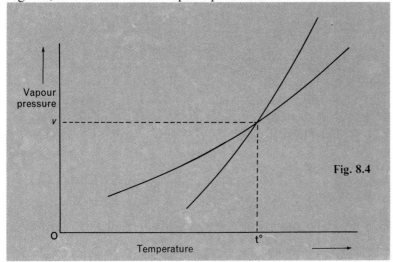

Fig. 8.4

The only logical *first* choice of oven temperature would be the safe maximum operating temperature of the stationary phase.

Return to ← page 170 and choose that alternative.

Your answer, $I = 626$, is wrong.

You have used the wrong values for n and z to arrive at this answer.

Remember our two alkanes were hexane (C_6H_{14}) and pentane (C_5H_{12}).

Derive n, where n is the difference in the number of carbon atoms for these n-alkanes. Now derive z, where z is the number of carbon atoms of the alkane having the lower number of carbon atoms.

Use these values to re-calculate I.

Now return and choose the correct alternative from those listed on → page 203.

You think that component A was identified as *n*-octane. Sorry, but you're wrong. Your answer shows that you have used 'absolute' retention times in the calculation and not the retention times relative to the air peak, for component A and the standard, *iso*-octane.

Remember, the formula for the *relative retention time* was

$$T_{tj} = \frac{(t - x)}{(j - x)}$$

where in this case t is the retention time of component A, i.e. 15·5 mm; j is the retention time of the standard, *iso*-octane, i.e. 47 mm; and x is the retention time of the air peak, i.e. 2 mm.

Substitute these values in the formula and calculate the *relative retention time* for component A. Your answer should not be 0·33.

Repeat the calculation for components B and C (their retention times were 17 mm and 62 mm respectively), then go back to → page 206 and choose the correct answer from those given.

back ref. page 203

Your answer, $I = 613$, is wrong.

To arrive at this answer, you have used the wrong value of z. Remember that z is the number of carbon atoms of the alkane having the lower number of carbon atoms. In our case this alkane is pentane (C_5H_{12}).

What should be the value of z then?

Use this value to re-calculate I.

Now return and choose the correct alternative from those listed on → page 203.

Your choice, a long column, would not produce the chromato-gram as quickly as possible.

Both nonane (b.p. 150·8°C) and *cis*-decalin (b.p. 195°C) have quite high boiling points, so they are of LOW volatility. Even at 75°C, the temperature of the analysis, their vapour pressures would be quite low. What effect would this have on their retention time?

Would a long column increase or decrease the effect?

Think about the answers to these questions, then return to ← page 174 and choose a better alternative.

Your answer, the peak widths (i.e. LN and TV), is wrong.

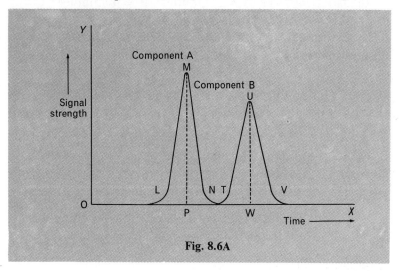

Fig. 8.6A

The peak width of each component measures the efficiency with which the column separates that particular component from the mixture, and unless the column is overloaded it will stay constant.

If the width stays constant, what is the only other parameter that can change when the amount of material in the detector changes?

Return to ← page 176 and choose the correct alternative.

Your answer, that the retention time of each component relative to the air peak was A = 2 minutes; B = 4 minutes; C = 7 minutes, is correct, for we can write:

Retention time of any component = Retention time of the air peak + Retention time relative to the air peak of that

component

If x = retention time of the air peak and y = retention time of a component relative to the air peak, then the retention time of any component = $x + y$.

Let n be the number of theoretical plates in the column. We can now define a number 'N'*, called the EFFECTIVE NUMBER OF THEORETICAL PLATES in the column as:

$$N = n \left(\frac{\text{retention time relative to the air peak of a component}}{\text{retention time of that component}} \right)^2$$

or $N = n \left(\dfrac{y}{x + y} \right)^2$

Question

In an experiment to find the effective number of theoretical plates in a particular column, the retention time of the air peak was found to be 2 minutes and that of a single substance 4 minutes. If the number of theoretical plates was calculated (from retention time and peak width at its base) to be 500, would your calculations show that the effective number of theoretical plates was:

1. 250?
2. 225?
3. 125?
4. none of these?

← page 158
← page 171
← page 173
← page 162

---

* L. S. Ettre, *J. Gas Chromatog.*, **1** (Feb.), 38 (1963).

Your answer, $I = 526$, is wrong.

By arriving at this answer you show that you have used the wrong value for n, which is the difference in the number of carbon atoms for the n-alkanes. Our alkanes were pentane (C_5H_{12}) and hexane (C_6H_{14}). What is the value of n?

Use this value to recalculate I.

Then return and choose the correct alternative from those listed on → page 203.

Right. If the amount of the stationary phase is increased, the amount of the component which it can dissolve will increase. This will lengthen the chromatographic process, and hence the retention time of that component.

As an approximate guide only to the amount of stationary phase to use when making up a column,

(i) Make up (or take up) a column containing 10 *w/w*% of the chosen stationary phase.

(ii) By gas–liquid chromatography find the value of x, the retention time of the air peak, and y, the retention time of the component, relative to that of the air peak.

Then, if $y > 7x$, use less stationary phase (i.e. down to a lower limit of about 5%).

If $y \to x$, use more stationary phase (i.e. up to about 30 *w/w*%).

If y lies between x and about $7x$, then the quantity of stationary phase that was used (i.e. 10 *w/w*%) is probably adequate.

Remember, IF YOU CHANGE THE TEMPERATURE, THE RETENTION TIMES WILL CHANGE TOO.

The optimum quantity of stationary phase will be found by experiment, after taking into consideration the effect of the temperature of the analysis.

Question

A mixture of two components was chromatographed on a column containing 10 *w/w*% of stationary phase, at constant temperature and constant volume flow rate of carrier gas. If, under these conditions, their retention times were 3 minutes and 3·5 minutes, and the retention time of the air peak was 1 minute, would you say that:

1. a column with less stationary phase should be used? → page 191
2. a column with more stationary phase should be used? → page 194
3. the amount of stationary phase is probably adequate? ← page 174

or Are you undecided about the right amount of stationary phase to be used? ← page 168

You don't think it wise to try and approximate the area. Full marks for being frank about it.

Of course, for the most accurate work the areas LMN and TUV must be measured precisely, but for rapid quantitative checks an approximation will do.

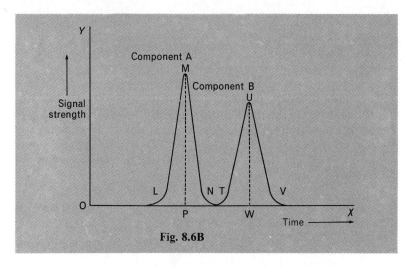

Fig. 8.6B

The peak width of each component measures the efficiency with which the column separates that particular component from the mixture and, unless the column is overloaded, will stay constant.

If the width stays constant, what is the only other parameter that can change when the amount of material in the detector changes?

Return to ← page 176 and choose the correct alternative.

Wrong. Look at the table again.

Compound	Vapour pressures at	
	75°C	150°C
Nonane	80 mm	760 mm
cis-Decalin	25 mm	400 mm

The vapour pressure of each component increases as the temperature is raised. If the vapour pressure increases would you expect the chromatographic process to speed up or slow down?

What effect would such a change have on the retention time of each component?

Think about the answers to these questions, then turn again to ← page 165 and choose a better alternative.

You think that cyclohexanone would be the best standard to use. No. Remember the standard should preferably lie about half way between our two components, that is, it should have a Kovats index between those of our two components, n-octane at 800 and ethylbenzene 1176.

Clearly cyclohexanone whose Kovats index is 1361 does not fulfil this requirement.

Return to → page 200 and choose the alternative standard which does satisfy the requirement.

Yes, you have chosen correctly, for the peak width of each component measures the efficiency with which the column separates that particular component from the mixture, and *at constant temperature* will stay constant, unless the column is overloaded. (If each peak width were plotted against the retention time for that peak *at constant temperature*, the resulting graph would be a straight line.)

Thus, when the amount of material in the detector changes, the peak height changes. Obviously this approximation is not valid if the peak is misshapen or very squat compared with others in the chromatogram.

As a *general approximation* then we can say that the *amount of any component present* ≡ *maximum peak height for that component*.

Now let us add a third component, C, to our mixture of components A and B. Assume that all other conditions are kept constant; then our new chromatogram could look like Fig. 8.7, where the retention time for component C is *f* minutes and its maximum peak height FH represents the quantity of component C in the mixture.

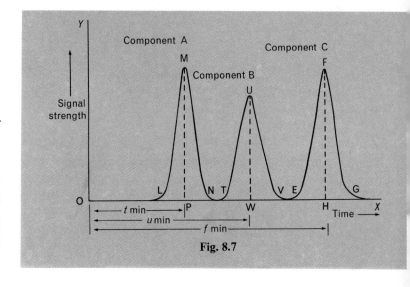

Fig. 8.7

Question

If component C had been added to the mixture because it was thought that the components B and C were identical, would this chromatogram have proved that

1. they were, in fact, identical? ← page 169
2. they were *not* identical? → page 198

If you think that it doesn't prove anything, turn to → page 192

You think a column with less stationary phase should be used. Sorry, but you're wrong.

You were told that the retention time of the air peak was 1 minute and the retention times of the two components were 3 minutes and 3·5 minutes. Calculate the values of y for each peak, i.e. the retention time of that peak relative to that of the air peak.

Remember we said that x was the retention time of the air peak, and if

> $y > 7x$ less stationary phase should be used
> > (i.e. down to about 5 $w/w\%$)
> $y \rightarrow x$ more stationary phase should be used
> > (i.e. up to about 30 $w/w\%$)
> y lies between x and $7x$ the amount of stationary phase was probably adequate.

Where do your calculated values of y lie?

Return to ← page 186 and select the correct alternative.

You can't see that the chromatogram proves anything. I'm afraid you're wrong. Here it is again (Fig. 8.7B).

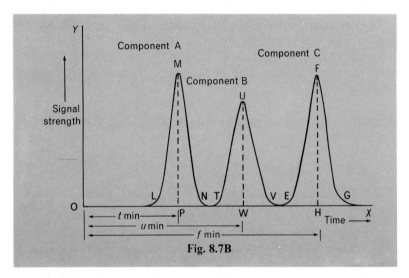

Fig. 8.7B

Here are the things it proves:

(i) Since the peak height of component **B** has not changed, the amount of component **B** in the detector is the same as for the previous analysis ← **page 176.**

(ii) Since the chromatogram shows three separate peaks with retention times $t, u,$ and f minutes the mixture injected contains three components.

Using these facts, go back to ← **page 190** and choose the correct alternative from the other two listed there.

Right. Here are the two chromatograms side by side.

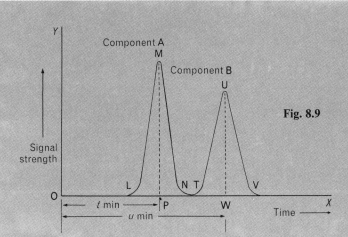

Fig. 8.9

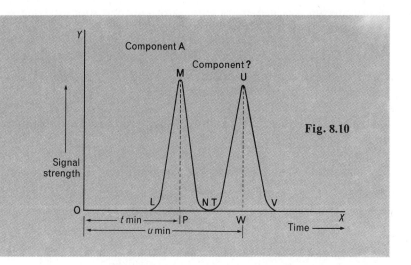

Fig. 8.10

Since in Figure 8.10 there are only two peaks whose retention times (t minutes and u minutes) are identical to those in Fig. 8.9, we can say that component C corresponds in retention time to either component A or component B. *But* since the peak height of component '?' shows a relatively strong increase as compared with the peak height for component B in Fig. 8.9 and their retention times are the same, we can say that components B and C are possibly identical. *This procedure then gives us a quick method for the qualitative analysis of a liquid mixture.*

Question

Four injections were made of an unknown mixture. Three peaks were obtained on each chromatogram and their retention times and peak heights are quoted in the table opposite. Under the same conditions the retention times of *n*-octane and ethylbenzene were 13·2 mm and 58·1 mm respectively. A little ethylbenzene was added to the unknown mixture and a new chromatogram obtained. It too contained three peaks, one of

Injection Number	Peak Heights (mm)			Retention Times (mm)		
	Peak A	Peak B	Peak C	Peak A	Peak B	Peak C
1	240	140	168	13	38	58
2	236	139	165	13·5	38·2	58·2
3	242	141	169	12·9	37·9	58
4	241	140	168	13·2	38·1	58·1

which was 175 mm in height whilst the other two peaks were about the same height as before. Which component was identified as possibly ethylbenzene?

1. A.   → page 204
2. B.   → page 202
3. C.   → page 206

Your answer, to use a column with more stationary phase, is wrong.

The retention time of the air peak (x) was 1 minute. The retention times of the two components, $(x + y_1)$ and $(x + y_2)$, were given as 3 minutes and 3·5 minutes respectively.

Calculate the values of y_1 and y_2, which are the retention times of each peak relative to our air peak.

Remember we said that x was the retention time of the air peak, and if

$y > 7x$ less stationary phase should be used,
    (i.e. down to about 5 $w/w\%$)
$y \rightarrow x$ more stationary phase should be used
    (i.e. up to about 30 $w/w\%$)
y lies between x and $7x$, the amount of stationary phase was probably adequate.

Where do your calculated values of y lie?

Return to ← page 186 and select the correct alternative.

Your choice is component B. Right. The relative retention times A = 13·5/45 = 0·3, B = 15/45 = 0·33 and C = 60/45 = 1·33. Since the quoted relative retention time is 0·33, then component B is almost certainly n-octane.

The reason why *relative retention times* are calculated from the retention time of each component relative to that of the air peak, is to eliminate any possible error due to gas hold-up in the column. All other errors due to column variables are eliminated in the dividing process, e.g., $(t - x)/(j - x)$.

In the quotations the standards used are selected from an internationally recognised list of internal standards drawn up by the IUPAC Committee*. Here they are:

b.p. °C	Compound	Retention Volumes at 160°C	
		Apiezon L	Carbowax 20M
−0·5	n-butane	1·9	0·6
80·1	benzene	11·4	10·2
138	p-xylene	40·0	24·8
79·6	methyl ethyl ketone	4·8	7·4
99·2	{ iso-octane or 2,2,4-trimethylpentane	4·3†	3·0†‡
210·8	naphthalene	0·82§	0·54§‖
161·1	cyclohexanol	35·0	64·1
155·65	cyclohexanone	37·5	54·7

'Preliminary Recommendations on Nomenclature and Presentation of Data in Gas Chromatography', see reprints of the Third Symposium on Gas Chromatography, Edinburgh, June 1960.

The choice of standard depends on (i) its retention volume (time), (ii) its chemical reactivity, e.g. it must be soluble in and not react with either the mixture to be analysed or the stationary phase.

Question

Bearing in mind all that has been said concerning the choice of stationary phase, do you think the retention time of any substance relative to a suitable standard would change if the stationary phase was changed?

1. Yes. → page 203
2. No. → page 199

† Retention relative to pentane at 100°C.
‡ Carbowax 1500 as stationary phase.
§ Retention relative to quinoline at 200°C.
‖ Polyoxyalkalene adipate as stationary phase.

You think that p-xylene would be the best standard to use. No, I'm afraid you are wrong.

Its Kovats Index I is 1180 which is very near to that of one of our components, ethylbenzene (Kovats Index 1176).

What would this mean in terms of the separation of these two components by g.l.c.?

Remember the standard should preferably lie about half way between our two components; this means, at a Kovats Index of approximately

$$\left[800 + \left(\frac{1176 - 800}{2} \right) \right].$$

Work out this value of I.

Then return to → page 200 and choose a better alternative.

You haven't been able to calculate I. Let's do it together. Here is the equation again:

$$I = 100 \left[n \left(\frac{\log R_x - \log R_z}{\log R_{z+n} - \log R_z} \right) + z \right]$$

In our experiment the standards were pentane (C_5H_{12}) and hexane (C_6H_{14}). Derive n, which is the difference in the number of carbon atoms for these n-alkanes.

Derive z, where z is the number of carbon atoms of the alkane having the lower number of carbon atoms.

The retention times were $R_x = 90$ mm, that of hexane $= 210$ mm, and that of pentane $= 79$ mm.

Now look up the logarithms of these retention times and substitute them in the factor

$$\left(\frac{\log R_x - \log R_z}{\log R_{z+n} - \log R_z} \right)$$

where R_z is the retention time of pentane and R_{z+n} is the retention time of hexane. Let the value of this factor be F. Now substitute these values in the formula for I which now becomes

$$I = 100 \left[n . F + z \right]$$

and calculate I.

Now go back to → page 203 and choose the correct alternative from those listed there.

You think that the chromatogram would prove that components B and C were *not* identical. Correct. The fact that under identical conditions three peaks have been obtained, each of different retention time, proves that no two components in the mixture are identical.

In fact, to overcome personal or experimental error, the analysis would be repeated several times. From the chromatograms so obtained, it would be found that each retention time could vary slightly; thus for component A the variation might be $\pm \delta t$. However, as long as the retention time of a peak fell within the limits $t \pm \delta t$ we could be sure that it was the peak for component A.

Question

Now suppose the chromatogram for our mixture of component A, B and C looked like this.

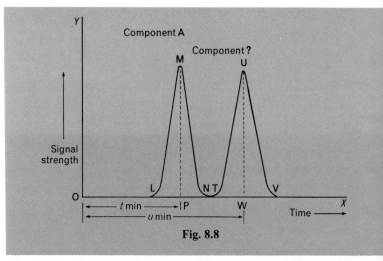

Fig. 8.8

Does this chromatogram (Fig. 8.8), assuming the conditions are identical again, prove that components B and C

1. are *possibly* identical?   ← page 19[3]
2. are *not* identical?   ← page 17[5]

If you are not sure what it proves.   → page 20[5]

Remember that we said that basically the process of chromatography depends on

(i) the partition coefficient, and hence the solubility of components in the stationary phase

(ii) chemical affinity and/or polarity effects between components and the stationary phase.

If the stationary phase changes then both these factors will change.

Return to ← page 195 and choose the other alternative.

You think it would be 513. Quite correct. Your equation for I looked like this:

$$I = 100 \left[1 \left(\frac{1.9542 - 1.8976}{2.3222 - 1.8976} \right) + 5 \right] = 513$$

To keep the retention index I as a whole number, the indices of the n-alkanes are defined as 100 times the number of carbon atoms in them. i.e. 100 for methane (CH_4); 200 for ethane (C_2H_6); etc. The index number above (i.e. 513) indicates immediately that *on that particular stationary phase* the unknown compound is eluted between pentane and hexane. *If the stationary phase is changed then the index number of the unknown also changes.**

Now we can see how the Kovats number can help us when choosing one of the IUPAC International Standards. As long as the stationary phase is the same, and the mixture is injected at the same time, then from the index number we can see in which order the components will be eluted, for that with the lowest number will be eluted first and so on and that with the highest number will be eluted last.

Let us look at an example. For the stationary phase Apiezon L at 160°C we find these values of I for the compounds listed:

Two-component mixture		IUPAC International Standards		
n-octane b.p. 126°C	ethyl-benzene b.p. 136°C	benzene b.p. 80°C	p-xylene b.p. 138°C	cyclo-hexanone b.p. 155.6°C
800	920	704	926	894

* A method of calculating I using a nomogram is given in Appendix A, p. 289.

Now, if we wished to measure the relative retention times of our two-component mixture of n-octane and ethylbenzene, using one of the three IUPAC Standards listed, from boiling point considerations we can see that the best one for our purpose would have a Kovats index between 800 and 920, namely cyclohexanone $(I = 894)$.

Question

The table below lists the Kovats indices for the stationary phase Carbowax 20M at 160°C.

Two-component mixture		IUPAC International Standards		
n-octane b.p. 126°C	ethyl-benzene b.p. 136°C	benzene b.p. 80°C	p-xylene b.p. 138°C	cyclo-hexanone b.p. 155.6°C
800	1176	979	1180	1361

Which standard would be the best choice for measuring the relative retention times of our two component mixture?

1. benzene,
2. p-xylene,
3. cyclohexanone,

→ page 207
← page 196
← page 189

back ref. page 206

You say that component C was *n*-octane. I'm afraid not.

The retention times were A = 15·5 mm; B = 17 mm; C = 62 mm; the retention time for *iso*-octane was 47 mm, and that of the air peak 2 mm.

A calculation of the retention time for component C relative to that of *iso*-octane shows that it is $\frac{(62-2)}{(47-2)} = \frac{60}{45} = 1\cdot33$.

Clearly a comparison with the relative retention time quoted in the literature of *n*-octane to *iso*-octane (i.e. 0·33) rules out component C as *n*-octane.

Now calculate the retention times of components A and B relative to *iso*-octane.

Which of these corresponds to 0·33?

Return to → page 206 and choose this alternative.

You have chosen component B. This is the wrong choice. Let's see why. Here is the data table again.

Injection Number	Peak Heights (mm)			Retention Times (mm)		
	Peak A	Peak B	Peak C	Peak A	Peak B	Peak C
1	240	140	168	13	38	58
2	236	139	165	13·5	38·2	58·2
3	242	141	169	12·9	37·9	58
4	241	140	168	13·2	38·1	58·1

After the addition of the ethylbenzene one peak height rises to 175 mm whilst the other two remain about the same height as before. Clearly from peak height considerations, component A is ruled out. The choice now lies between components B and C, as both their peak heights could rise to 175 mm in height.

Compare the average retention times (calculated from the table above) of components B and C with that of ethylbenzene which was 58·1 mm.

Which component has an average retention time of 58·15 mm?

Return to ← page 193 and select this alternative.

Correct. Because of differences in solubility, polarity effects etc., a change of stationary phase would result in a change in relative retention times.

To provide an easy means of identifying an unknown compound, Kovats* proposed an index system based on the *n*-alkanes as reference substances, since they are chemically inert, soluble in the common stationary phases and are non-polar.

The Kovats index, I, is calculated from the equation:

$$I = 100 \left[n \left(\frac{\log R_x - \log R_z}{\log R_{z+n} - \log R_z} \right) + z \right]$$

where R_x is the retention time of unknown substance X,
R_z is the retention time of the normal alkane having z carbon atoms,
R_{z+n} is the retention time of the normal alkane having $z + n$ carbon atoms.
and n is the difference in the number of carbon atoms for the *n*-alkanes.

Let's see how it works. A substance X was thought to have 7 carbon atoms. It was mixed with hexane (C_6H_{14}) and octane (C_8H_{18}) and a chromatogram was obtained, Fig. 8.13. $R_6 = 210$ mm; $R_8 = 795$ mm and $R_x = 470$ mm. $n = 8 - 6 = 2$; $z = 6$.

Then:
$$I = 100 \left[2 \left(\frac{2 \cdot 6721 - 2 \cdot 3222}{2 \cdot 9004 - 2 \cdot 3222} \right) + 6 \right] = 100 \left[2 \left(\frac{0 \cdot 3500}{0 \cdot 5782} \right) + 6 \right]$$

$$= 100 \left[1 \cdot 21 + 6 \right] = 721$$

* E. Kovats, *Helv. Chim. Acta.* **41**, 1951 (1958).

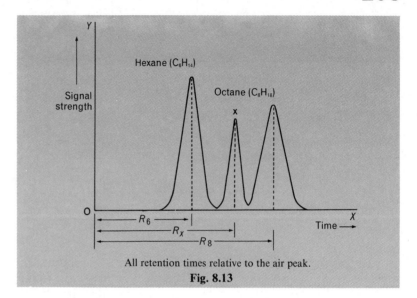

All retention times relative to the air peak.

Fig. 8.13

Question

In an experiment similar to that described, pentane (C_5H_{12}) and hexane (C_6H_{14}) were used as the standards. The retention times were $R_x = 90$ mm; $R_6 = 210$ mm; $R_5 = 79$ mm. What value have you calculated for the retention index?

1. 513   ← page 200
2. 526   ← page 185
3. 613   ← page 181
4. 626   ← page 179

If you haven't been able to calculate I,   ← page 197

Your choice, component A, is incorrect. Here is the data table again.

Injection Number	Peak Heights (mm)			Retention Times (mm)		
	Peak A	Peak B	Peak C	Peak A	Peak B	Peak C
1	240	140	168	13	38	58
2	236	139	165	13·5	38·2	58·2
3	242	141	169	12·9	37·9	58
4	241	140	168	13·2	38·1	58·1

Now after the addition of the ethylbenzene to the mixture, the new chromatogram contained 3 peaks one of which was 175 mm in height whilst the other two peaks were about the same height as before. But your choice, Peak A, was already 240 mm in height. Clearly then component A can't be ethylbenzene. The choice now lies between components B and C as both their peak heights could increase to 175 mm in height. To differentiate between them we look at each of their retention times and compare them with the retention time of ethylbenzene which was 58·1 mm.

Which component has an average retention time of 58·15 mm?

Return to ← page 193 and select this alternative.

You are unsure what the chromatogram proves. Let's look at it together.

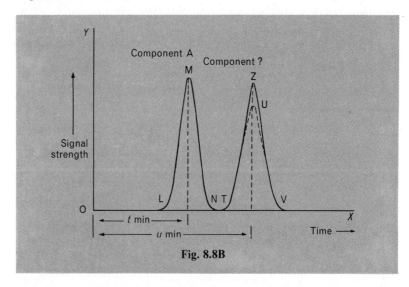

Fig. 8.8B

First, you will notice that there is a peak whose retention time of u minutes is equal to that of component B.

Secondly, the height of the peak whose retention time is u minutes shows an increase, whilst that of component A has remained the same. This must mean that there has been an increase in the concentration of this component in the mixture.

Since only component C was added, what do these facts prove about component B and component C?

Return to ← page 198 and choose the correct alternative.

You say it was component C. Quite correct. Only the peak heights of components B or C *could* rise to 175 mm, but from their retention times, only that of component C is, within experimental error, the same as that of ethylbenzene. These two facts together enable us to deduce that component C is possibly ethylbenzene.*

Literature quotations often refer to the '*relative retention times.*' These are calculated for the components of a mixture by choosing a central component D as standard. Let's do this for our mixture of components A and B. Here is the chromatogram again (Fig. 8.11):

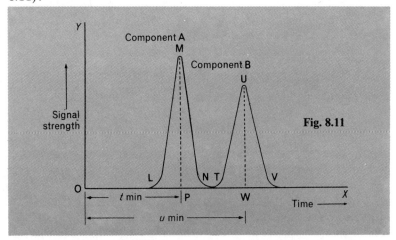

Fig. 8.11

We now choose component D whose peak comes in a 'space' in the chromatogram, centrally if possible, and add a little of D to our mixture of components A and B. A chromatogram is obtained, keeping the conditions constant as before, and will show three peaks (Fig. 8.12). The retention times, t minutes, u minutes and

Positive identification of the compound needs confirmatory evidence by other techniques. These are outlined in Part Eleven, Sect. 4.

j minutes, are measured. The retention time of the air peak under the same conditions was found to be x minutes. Designating the *relative retention time* for component A as T_{tj} and B as T_{uj}, we have $T_{tj} = \dfrac{(t - x)}{(j - x)}$, $T_{uj} = \dfrac{(u - x)}{(j - x)}$, and for component D,

$$T_{jj} = \frac{(j - x)}{(j - x)} = 1.$$

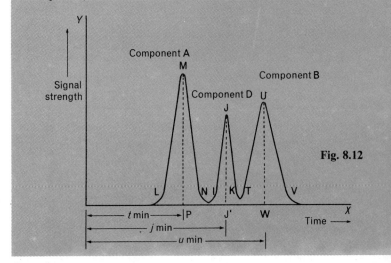

Fig. 8.12

Question:

In an analysis of a mixture of three components, A, B and C, suspected of containing *n*-octane, the retention times were as follows: A = 15·5 mm; B = 17 mm; C = 62 mm. *Iso*-octane as standard was added to the mixture and its retention time was 47 mm, whilst that of the air peak was 2 mm. Given that the retention time of *n*-octane relative to *iso*-octane was 0·33 which peak was identified as possibly due to *n*-octane?

1. A, ← page 180, or 2. B, ← page 195 or 3. C, ← page 201

This time you have selected benzene. Quite right. Its index of 979 shows that it will be eluted almost halfway between *n*-octane and ethylbenzene, whose indices are 800 and 1176.

As a general rule, then, the *n*-paraffins are useful as standards, except where:

(i) the appropriate paraffin is already present in the mixture or is not readily available, and

(ii) where other difficulties such as 'tailing' occur. (This will be discussed later in the programme.)

Part 9 starts on → page 208

Part Nine
Quantitative Analysis

WE CAN now apply the principles we have learned regarding the choice of internal standards, to the quantitative analysis of a mixture, for the same criteria apply, i.e.

(i)   The standard must be miscible with the sample to be analysed.
(ii) It must not react with any component(s) in the sample.
(iii) It should give only one peak and this should not overlap any sample component peaks.
(iv) Its retention time should be close to the sample component which is of interest (Kovats index will help in items iii and iv).

These criteria ensure the greatest accuracy in the analysis.

Question
If the mixture to be analysed contains compounds of interest which are eluted at very different times, say more than 15 minutes, do you think that the best accuracy could be maintained by:

1. using a standard whose peak is eluted about
   7 minutes after the first peak of interest?   → page 216
2. using two standards whose peaks fall close to
   those of the compounds of interest?   → page 221

**If you are not sure what to do to maintain the
accuracy.**   → page 224

You would construct a calibration graph for the expected region. This would not give the most accurate analysis.

Here is a portion of the curve above 30%. (Fig. 9.4A).

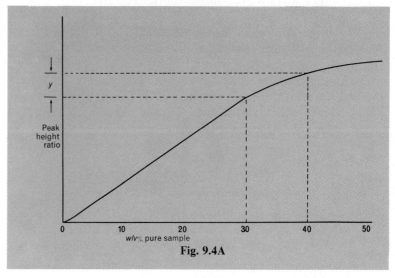

Fig. 9.4A

You will notice that for a small change in peak height ratio we have a large change in $w/v\%$ pure sample.

What effect would this have on the accuracy of the results above $30\ w/v\%$?

Return to → page 214 and choose a better alternative.

You said that $A_u/A_s = 0.09/0.164 = 0.55$. Right.

Now look at the chromatogram again (Fig. 9.2).

The two other ways we can measure the areas are:

(iii) Providing the peaks are symmetrical about the peak axis and of a shape similar to those shown in Fig. 9.2 (i.e. Gaussian), measure the peak heights BD $= h_u$ and FH $= h_s$, (where u and s signify 'unknown' and 'standard' peaks), and the width at half peak height i.e. UV $= w_u$ and WZ $= w_s$. Then the areas A_u and A_s are $A_u = K . h_u . w_u$, and $A_s = K . h_s . w_s$, and the ratio

$$\frac{A_u}{A_s} = \frac{h_u w_u}{h_s w_s},$$

where K is a constant.

(iv) If we draw the inflection tangents LM, NM and RS, ST, then designating area ABC as A and area LMN as A', these areas are linked by the equation $A = K'A'$ where K' is a constant.

If we write DM $= l_u$, HS $= l_s$, LN $= v_u$, and RT $= v_s$, then

$$\frac{A_u}{A_s} = \frac{l_u v_u}{l_s v_s}$$

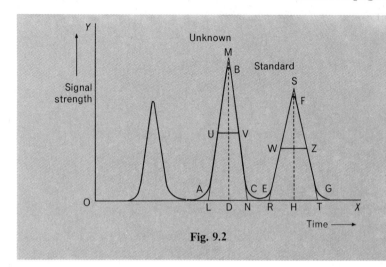

Fig. 9.2

Question

In a chromatogram the following measurements for the unknown and standard peaks were obtained: $h_u = 246$ mm; $h_s = 217$ mm; $l_u = 253$ mm; $l_s = 231$ mm; $w_u = 1.75$ mm; $w_s = 0.8$ mm; $v_u = 3.25$ mm; $v_s = 1.6$ mm. Calculate the peak area ratio to one place of decimals, using each formula given above. Which of the following values did you obtain?

1. 2·5 or 2·2 → page 223
2. 2·4 or 2·3 → page 215

If you don't understand how to calculate the peak area ratio. → page 226

Your answer, 0·3 $w/v\%$, is wrong.

You have calculated the peak height ratio correctly, but by choosing this alternative you have demonstrated that you have misread your calibration graph.

Take up your graph again and read off the % pure sample corresponding to a peak height ratio of 1·1.

Then return to → page 223 and choose the correct alternative.

Wrong. Remember the Peak Area ratio is

$$\frac{A_u}{A_s} = \frac{\text{Area of 'unknown' peak}}{\text{Area of 'standard' peak}} = \frac{\text{Weight of 'unknown' peak}}{\text{Weight of 'standard' peak}}$$

Your answer shows that you have confused these.

The values given were:

Weight of 'unknown' peak   0·09 g.
Weight of 'standard' peak   0·164 g.

Substitute these figures in the equation, and recalculate A_u/A_s.

Return to → page 221 and choose this alternative.

You have arrived at an answer which was neither of those listed. Let's try to see where you went wrong.

After the dilution we had x g of pure sample dissolved in $40 + 90 = 130$ ml of solvent. The graph showed that the strength of this solution was 16 $w/v\%$. We can find x from these figures thus:

$$x = \frac{16}{100} \times 130 \text{ g.}$$

Calculate x yourself. But originally this amount, (x g) of pure sample was dissolved in only 40 ml of solvent. The strength of the original solution must have been

$$\frac{x}{40} \times 100 \; w/v\% = y \; w/v\%$$

Calculate y yourself. Did you see where you made your mistake?

Turn again to → page 225 and choose the correct alternative from those listed there.

214

Your answer, 4·3 $w/v\%$, obtained from the peak height ratio of 9·4/8·6 = 1·1, is correct.

Since in the examples given the solutions have been fairly dilute, the calibration graph has been linear. If, however, we were to plot the graph for the full range from very dilute solutions to concentrated solutions it would not be linear below 1 $w/v\%$ and above about 30 $w/v\%$.

If the 'unknown' solution was dilute, then the calibration curve must be plotted in more detail over the lower range. This means adding to the standard, amounts of pure sample of say 0·05, 0·1, 0·2, 0·5, 0·7 and 1·0 $w/v\%$. This would give a calibration graph like Fig. 9.4.

As before, a peak height ratio of the 'unknown' of value y, corresponds to a value of 0·8 $w/v\%$ pure sample in the 'unknown'.

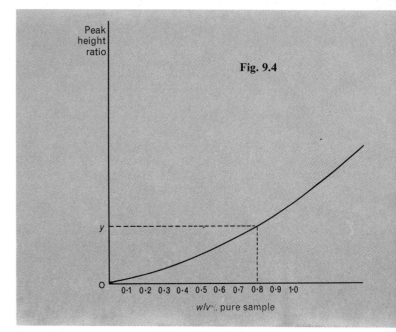

Fig. 9.4

Question

Suppose the 'unknown' mixture contained more than 30 $w/v\%$ of pure sample. To analyse it quantitatively, would you:

1. construct a calibration graph for the expected region, similar to that plotted for very dilute solutions? ← page 209

2. dilute the concentrated solution with a known volume of solvent, until the subsequent points on the calibration graph fell in the linear part of the graph? → page 225

If you would choose neither of these alternatives. → page 217

You have calculated that the peak area ratio lies between 2·3 and 2·4. Well, it may do, since these limits are within the correct limits of 2·2 and 2·5, but on the other hand it may not.

By arriving at this answer you have shown that you have confused peak height h with the height of the triangle l.

Here is a Gaussian peak ABC with the inflection tangents LM and NM drawn in. The peak height = BD = h. The height of the triangle LMN = MD = l.

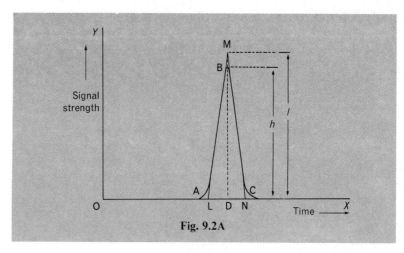

Fig. 9.2A

Return to ← page 210 and recalculate the peak area ratio using the two formulae $\dfrac{A_u}{A_s} = \dfrac{h_u \cdot w_u}{h_s \cdot w_s} = \dfrac{l_u \cdot v_u}{l_s \cdot v_s}$ and choose the correct alternative.

By using a standard whose peak is eluted about 7 minutes (about half way) after the first peak of interest, the best accuracy would not be maintained.

Remember, we said that one of the principles regarding the choice of an internal standard was that *its retention time should be close to the sample component which is of interest.*

Now turn to ← **page 208 and choose an answer which fits this clue.**

You would choose neither of the alternatives. Fine. If you have a good idea pass it on to your supervisor.

I hope you have chosen this alternative because you can see that alternative **1** is not the best, because as the curve flattens out above 30 % the accuracy becomes very poor.

But can you see that by diluting the sample with more solvent the concentration of pure sample would decrease, bringing the peak height ratio on to the linear portion of the graph?

Return to ← page 214 and choose this alternative.

You have found that the amount of pure component is 3·5 $w/v\%$. That's wrong, I'm afraid.

Your answer shows that you have miscalculated the peak height ratio. Remember it is the ratio

$$\frac{\text{peak height of pure component}}{\text{peak height of standard}}$$

Recalculate the peak height ratio of your 'unknown'; it is $\dfrac{9\cdot4}{8\cdot6}$

Now from your calibration graph read off the % pure sample in the solution which would give this ratio.

Then return to → page 223 and choose the correct alternative.

Your answer was none of those listed. Let's see where you went wrong.

The Peak Area ratio is:

$$\frac{A_u}{A_s} = \frac{\text{Area of 'unknown' peak}}{\text{Area of 'standard' peak}} = \frac{\text{Weight of 'unknown' peak}}{\text{Weight of 'standard' peak}}$$

The values given were:

Weight of 'unknown' peak   0·09 g.
Weight of 'standard' peak   0·164 g.

Then $\dfrac{A_u}{A_s} = \dfrac{0·09}{0·164}$

Work this out, and select the appropriate alternative on → page 221

Your answer was none of those listed. Let's see if you calculated the peak height ratios correctly. Here is the table again:

$w/v\%$ pure component	Peak height (cm) of pure component	Peak height (cm) of standard	Peak height ratio
0·5	1·3	8·6	1·3/8·6 = 0·15
2	4·7	8·7	4·7/8·7 =
5	10·7	8·7	10·7/8·7 = 1·23
7·5	16·6	8·5	16·6/8·5 =
10	21·9	8·6	21·9/8·6 = 2·54

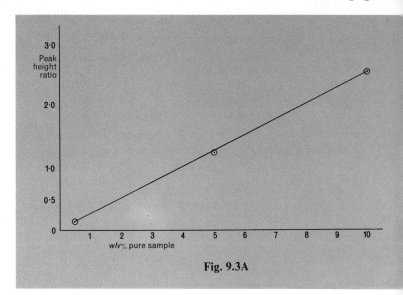

Fig. 9.3A

Fill in the blanks in the table.

Now let's see if you have drawn the calibration graph correctly It should look like Fig. 9.3A, (if it does not, then try to see where you went wrong).

Using this graph, try the question again. It was, that on analysing a mixture containing an unknown amount of pure component, the peak height for that component was 9·4 cm and that for the standard 8·6 cm. Calculate the peak height ratio and use the graph to find the amount of pure component in the mixture.

Return now to → page 223 and choose the correct alternative.

Right. In this situation we would use two standards whose peaks all close to those of the compounds of interest.

n the most precise work a standard solution of the internal tandard is made up in the solvent. A fixed volume of this olution (say x ml) is added to the mixture and then more solvent ldded to bring the total volume to y ml. A sample of this solution s then chromatographed giving a series of peaks, including the peak from the standard added. Suppose the chromatogram ooked like Fig. 9.1.

We know that areas ABC = A_u and EFG = A_s represent the amount of 'unknown' and 'standard' materials present in the mixture. These areas can be measured in four ways. Here are wo of them.

i) They can be measured using a planimeter, or on some g.l.c. machines they are measured automatically by an integrator.

ii) We can cut out the areas from the chromatogram with scissors and weigh the paper cut-outs accurately. Then the peak area ratio A_u/A_s is obtained by simple division.

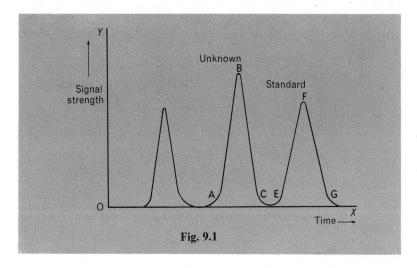

Fig. 9.1

Question

An internal standard was made up in ether solution and x ml of it was added to an unknown mixture. Ether was added until the volume was y ml. An injection of $1\,\mu$l of this solution was made and the chromatogram contained two peaks, which were cut out and weighed. Here are the results: Weight of 'unknown' peak was $0.09\,$g; weight of 'standard' peak was $0.164\,$g. From these figures what is the peak area ratio A_u/A_s?

1. 0.55   ← page 210
2. 1.82   ← page 212
3. neither of these figures.   ← page 219

back ref. page 225

You have calculated that the strength of the original solution was 36 $w/v\%$. Sorry, but you're wrong.

By arriving at this answer you have shown that you have forgotten to include the original 40 ml of solvent when calculating the number of grams of pure sample in the solvent. The total amount of solvent is $90 + 40 = 130$ ml

Use this figure to recalculate the strength of the solution.

Then return to → page 225 and choose the correct alternative.

Your answer was that the peak area ratio A_u/A_s was 2·5 or 2·2.
Right.

This ratio tells us that there is more than twice the quantity of one component than the other in the mixture. You will readily see that although the peak heights are nearly the same for each component, their base widths are in the ratio of over 2:1. Although measurement of peak areas is the most accurate way of analyzing a mixture quantitatively, for most normal purposes it is only necessary to measure the peak height ratio, since the peak height due to any substance increases very nearly linearly with increase in the amount of the component provided that: (i) the peaks are of similar shape, (ii) their base widths are almost equal.

We then have, $\dfrac{A_u}{A_s} \simeq \dfrac{h_u}{h_s}$ = Peak Height Ratio

Let's see how this works in practice. We take say, five samples of the standard solution each of x ml. To them we add samples of pure component so that when each of the five samples is made up to a fixed volume (y ml) with the solvent, the weight/volume percentage of the pure sample would be say 10, 7, 5, 2, 1.

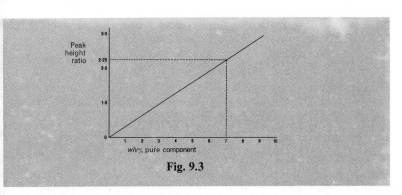

Fig. 9.3

Each solution is now chromatographed, the peak heights measured and the peak height ratio calculated. This ratio is then plotted against the $w/v\%$ of the pure component in each solution giving a *calibration graph* which would look something like Fig. 9.3.
Now if the solution of unknown strength were treated similarly and the peak height ratio was found to be say, 2·25, then from this graph we can see that the solution contains exactly 7 $w/v\%$ of pure component.

Question
In an experiment similar to that described the following results were obtained:

$w/v\%$ of pure component	Peak Height (cm) of pure component	Peak Height (cm) of standard
0·5	1·3	8·6
2	4·7	8·7
5	10·7	8·7
7·5	16·6	8·5
10	21·9	8·6

Calculate the peak height ratios and draw the calibration graph on squared paper. On similarly analysing a mixture containing an unknown amount of pure component, the peak height for that component was 9·4 cm and that for the standard 8·6 cm. From this information, using your calibration graph, what was the amount of pure component?

1. 0·3 $w/v\%$ ← page 211   3. 4·3 $w/v\%$ ← page 214
2. 3·5 $w/v\%$ ← page 218   4. None of these.   ← page 220

You're not sure what to do to maintain the accuracy. The clue lay in the criteria listed.

Provided that the standard is miscible with the sample and does not react with it, the standard should give only one peak which does not overlap any sample component peak. Its retention time should be close to the component which is of interest.

Now turn again to ← page 208 and choose the alternative which fits all these conditions.

Correct. By diluting the solution with more solvent, a peak height ratio which fell on the linear portion of the graph would be obtained. Suppose that this ratio corresponded to a content of 5 $w/v\%$ and that to the original volume of the solution of 10 ml, 70 ml of solvent had been added. Then the weight of the pure sample is $\dfrac{5}{100} \times 80\,g = 4\,g$.

But this 4 g was originally dissolved in only 10 ml of solvent. *Thus the strength of the original unknown solution* was

$$\frac{4}{10} \times 100\ w/v\% = 40\ w/v\%$$

Question

To analyse quantitatively a solution containing x g of pure sample in 40 ml of solvent, it was necessary to dilute the solution with another 90 ml of solvent before the peak height ratio fell on the linear portion of the calibration graph. This peak height ratio corresponded to 16 $w/v\%$ of pure sample in the solution. From these facts what value do you obtain for the strength of the original solution?

1. 36 $w/v\%$ ← page 222
2. 52 $w/v\%$ → page 228
3. neither of these.   ← page 213

You're not sure you understand how to calculate the peak area ratio. Well let's work it out. Look again at the chromatogram Fig. 9.2B.

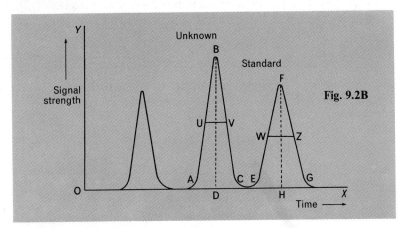

Fig. 9.2B

There are two ways in which the peak area ratio can be calculated The first way depends on the fact that the area of a Gaussian peak (that is a peak of the same shape as those produced in our chromatogram), let's call it A, is equal to a constant (K) multiplied by the peak height (h) and the peak width at half peak height (w). We can write this as

$$A = K.h.w.$$

As long as the peaks are Gaussian in shape, K is the same for each peak. Writing the area of the 'unknown' peak as A_u, the peak height as h_u and the peak width at half height as w_u, we can write $A_u = Kh_uw_u$. Similarly for the 'standard' peak $A_s = Kh_sw_s$. *The peak area ratio then becomes*

$$\frac{A_u}{A_s} = \frac{Kh_uw_u}{Kh_sw_s} = \frac{h_uw_u}{Kh_sw_s}$$

Now look at the chromatogram again

$$h_u = BD; \quad w_u = UV; \quad h_s = FH; \quad w_s = WZ.$$

and we were given the values $h_u = 246\,mm$; $w_u = 1.75\,mm$; $h_s = 217\,mm$; $w_s = 0.8\,mm$. Substituting in our formula we have

$$Peak\ Area\ Ratio = \frac{A_u}{A_s} = \frac{(246)(1.75)}{(217)(0.8)}$$

Now calculate the peak area ratio, then turn to → page 227

Your answer should be 2·5. Right. Here's the chromatogram again (Fig. 9.2C). The second way in which the peak area ratio can be calculated is as follows. If we draw the inflection tangents to a Gaussian curve (our peak is Gaussian, remember), e.g. for our unknown peak these are LM and NM, then it can be proved that the relationship between the actual area of the peak ABC(= A) and the area of the triangle LMN(= A') is given by the equation $A = K'A'$ where K' is a constant which is the same for peaks of similar Gaussian shape. Now we know that the area of triangle LMN = $\frac{1}{2}$(LN)(MD) = A'_u.

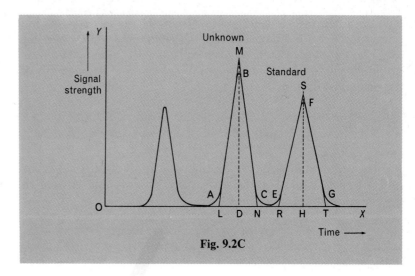

Fig. 9.2C

Using the symbols as before where l_u = DM; l_s = HS; v_u = LN; and v_s = RT,

$$A_u = K'A'_u = \frac{K'v_u l_u}{2}$$

Similarly
$$A_s = K'A'_s = \frac{K'v_s l_s}{2}$$

The Peak Area Ratio then becomes $\dfrac{A_u}{A_s} = \dfrac{v_u l_u}{v_s l_s}$

We were given that l_u = 253 mm, l_s = 231 mm, v_u = 3·25 mm and v_s = 1·6 mm. Substituting in our formula we have

$$\textit{Peak Area Ratio} = \frac{A_u}{A_s} = \frac{(3·25)(253)}{(1·6)(231)}$$

Calculate this peak area ratio and then return to ← page 210 and choose the alternative which lies between these limits.

You obtained the value 52 $w/v\%$. Right. There was 16 $w/v\%$ of pure sample in 130 ml of solution, thus there was $16/100 \times 130$ g $= 20.8$ g of pure sample in the solvent. But this 20.8 g was originally dissolved in only 40 ml of solvent, so the strength of the original solution was

$$\frac{20.8}{40} \times 100 = 52 \, w/v\%$$

It is as well to note that these principles apply whether it is the ratio of *peak heights* or the ratio of *peak areas* that is used to draw the calibration graph.

Part 10 starts on → page 229

Part Ten

Interpretation of the Chromatogram[*]

What has been said up to now has assumed that everything has gone smoothly and a perfect chromatogram has been obtained in each case. It would be naïve to expect this to happen always in practice.

One factor which affects the retention time of any peak, is the pressure of the carrier gas. In a gas chromatographic column the particles of packing offer resistance to the flow of carrier gas because of the finite viscosity of the gas and so there is a pressure gradient along the length of the column. This gradient would have no effect upon retention volume and thus retention time, were it not for the fact that the gas is compressible. Because of this the volume flow rate is greater at the outlet than the inlet. Thus a velocity gradient is inevitable when there is a pressure gradient in the column.

Question

Which of the following factors would NOT help to minimise the velocity gradient?

1. coarse packing   → page 246
2. a longer column   → page 235
3. slow flow rate of carrier gas   → page 240

[*] The chromatograms in this part of the programme are reproduced by kind permission of Perkin–Elmer Ltd., Beaconsfield, Bucks.

No. You were told that the analysis time was reasonable. It is the resolution that is poor. Reducing the carrier gas pressure would not help the resolution here. It is the selectivity of the chromatographic process that is at fault.

Which of the suggestions would most improve the chromatographic process?

If you do not know, re-read ← page 74, then try to make the correct choice on → page 237.

back ref. page 250

You would lower the oven temperature? Correct. This would rectify the peak asymmetry caused by thermal decomposition of the sample. An increase in the volume flow rate of the carrier gas and a reduction in the amount of stationary phase, down to about 3 $w/w\%$ would both help by shortening the elution times. (Note also that if an injection is made into a hot zone, the temperature of this zone might have to be reduced, to prevent thermal decomposition).

Asymmetry of peaks should not be confused with the 'tailing' of peaks which looks like Fig. 10.5.

This may not interfere with the analysis, and can often be eliminated by using a tailing inhibitor, and/or using a column containing a more inert support material.

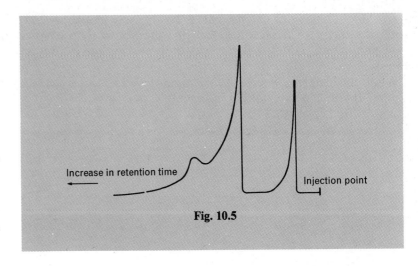

Fig. 10.5

Question
Another cause of tailing could be:

1. temperature too high
2. using the wrong stationary phase
3. injecting too much sample.

→ page 236
→ page 244
→ page 248

A reduction in the carrier gas pressure may make a slight improvement in the resolution, if it were above the optimum originally, but there is a better way of improving the resolution.

What effect does a decrease in temperature have on retention time? If you can't remember look again at → **page 235**

In the light of your answer to this question go back to → page 242 and make a better choice from those listed.

You would increase the column temperature by 60°C to 140°C. No. Remember we said that the rule was AS TEMPERATURE RISES, COLUMN EFFICIENCY DECREASES.

What do we need to do, regarding column efficiency, to effect a separation between *m*-xylene (b.p. 139°C) and *p*-xylene (b.p. 138°C)?

Think about this, then return to → page 235 and choose a better alternative.

back ref. page 244

Right. An increase in the column temperature would cause an increase in the vapour pressure of the components passing through the column. This would reduce their solubility in the stationary phase and cause them to be eluted faster. An increase in the volume flow rate of the carrier gas would reduce the elution time still further, so taking both actions together would be the quickest way to shorten the retention times and hence the total analysis time. One must, however, be careful about thermal decomposition if the temperature is raised too much.

Another result of the vapour pressure increase, caused by raising the column temperature, would be that the height of the peaks in the chromatogram would increase relative to their width.

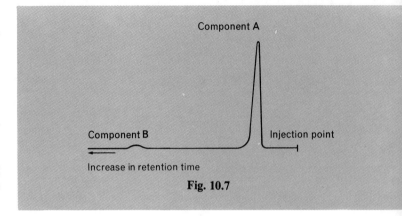

Fig. 10.7

Question
Now look at the above chromatogram. Would you say that

1. The analysis time is adequate, but the peak from component B is too small for useful interpretation? → page 254
2. the analysis time is too long and the peak from component B is too small for useful interpretation? → page 269
3. the analysis time is too long but the peak from component B is large enough to be interpreted? → page 267

Your choice, a longer column, was quite correct. The velocity gradient can be minimised by the use of coarse packings, slow flow rates and short columns.

It can be shown that a large pressure ratio between the inlet and outlet of the column (P_i/P_o) has an adverse effect on the performance of the column and it is desirable to design columns so that this pressure ratio approaches unity. This is why columns work better with pressures in excess of atmospheric rather than under vacuum.

A second factor which affects an analysis is the effect of temperature. A change in temperature affects (i) the partition coefficient, (ii) the gas flow rate in the column for a given pressure ratio P_i/P_o.

As a result of both these effects it can be shown that relative retention times are greater at low temperatures than high temperatures and thus reduction in temperature increases column efficiency. On the other hand the use of low temperatures results in long analysis times and it is easier to overload the column, so small samples must be injected. The choice of a suitable operating temperature, then, depends on balancing these factors to give the desired result.

Question

A chromatographic analysis of a mixture of *m*-xylene (b.p. 139°C) and *p*-xylene (b.p. 138°C) was carried out using a semi-polar stationary phase, di-nonyl phthalate at 80°C. No separation was obtained. Which of the following courses would you adopt to effect a separation?

1. increase the column temperature to about 140°C.   ← page 233
2. decrease the column temperature to 20°C.   → page 241
3. change the stationary phase.   → page 255

If you think that none of these courses would effect a separation.   → page 245

You think that 'tailing' is caused by too high a temperature. No. In fact the converse is true, tailing can result from too low a temperature causing undue condensation of the sample.

If the temperature was high enough to cause the decomposition of the sample then asymmetrical peaks could be produced in the chromatogram.

Are you confusing 'tailing' and asymmetry?

Return to ← page 231 and study the shape of those peaks in comparison to those on → page 250 then try to select a better answer.

You would reduce the column temperature. Quite correct. Since a reduction in column temperature would

i) change the partition coefficient,
ii) lower the gas flow rate in the column, for a given P_i/P_o,

the effect would be to lengthen the retention times of the components considerably. The separation of the peaks would increase, so we would have better resolution.

Here is a chromatogram which could have been obtained from a first injection (Fig. 10.2.)

The resolution is poor and the peaks are barely discernable, although the analysis time and retention times are reasonable. One cause of poor resolution is poor selectivity of the chromatographic process.

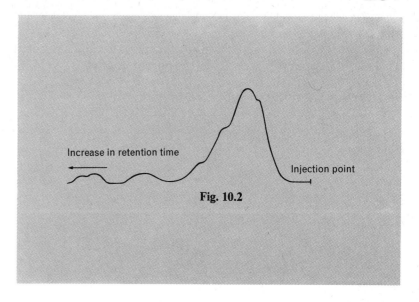

Increase in retention time

Injection point

Fig. 10.2

Question
In order to improve the resolution, by increasing the separation of the peaks in this case, would you:

1. reduce the carrier gas pressure? ← page 230
2. reduce the column temperature? → page 247
or 3. use a column with a different stationary phase? → page 257

back ref. page 257

No. I'm afraid you are wrong.

What effect would an increase in the quantity of sample injected have on the areas of the peaks in the chromatogram?

Would this improve the resolution or not?

Think about the answers to these questions then choose a better alternative on → page 257.

You would do none of the things listed to rectify the thermal decomposition of the sample. Well, if you know of another way, I'll be pleased to hear of it.

Let's consider the suggestions in turn:

(i) *Lower the oven temperature*: Would you expect that *thermal* decomposition would be reduced or cease altogether if this course were adopted?

(ii) *Increase the flow rate of the carrier gas.*   and

(iii) *Decrease the amount of stationary phase.* Both actions (ii) and (iii) would tend to shorten the retention times of the components in the mixture. What effect would this have on the contact time between the components and the hot column?

Would actions (ii) and/or (iii) *completely* rectify the thermal decomposition of the sample?

Think about the answers to these questions, then return to → page 250 and make the best choice from those listed there.

No, a slow flow rate of carrier gas *would reduce* the velocity gradient through the column.

Remember the question asked for the alternative that would NOT help to minimise the velocity gradient.

Return to ← page 229 and choose a better alternative.

Correct. Relative retention times are greater at lower temperatures and column efficiency increases. Since in this case we can ignore polarity effects, lowering the temperature to 20°C effects the separation.* The retention times proved to be about 25 to 30 minutes and the separation factor 1·05.

There may well be other selective stationary phases which would be better than di-nonyl phthalate in effecting a separation between *m*-xylene and *p*-xylene (e.g. cresyl phosphate). However, once a stationary phase has been chosen, the phase would be changed only after lowering the temperature as far as possible had failed to separate the components.

Before we go into the detailed interpretation of a chromatogram here is a brief general summary of the factors which will affect retention time.

No.	Factor	Retention times Short	Retention times Long
1	Volume flow rate of the carrier gas	high	low
2	Concentration of stationary phase on the support material	low	high
3	Solubility of component in the stationary phase	low	high
4	Volatility	high	low
5	Temperature (oven)	high	low
6	Length of column	short	long

Consider these factors carefully, then turn to → page 242.

---

* W. A. Wiseman, *Nature*, **185**, 841 (1960).

Usually some information about a sample is available before analysis and from this, approximate settings of the instrument controls can be chosen. After seeing the chromatogram from the first injection, it may be necessary to alter any or all of the control settings several times, until a satisfactory result is obtained.

One practical point. Up to now chromatograms have been shown with the peaks appearing from left to right. In fact, due to the way some recorders work, the peaks can appear in order from right to left. *As a general rule, the injection point will be near the end having a region of narrow peaks, close together.*

Here is a chromatogram (Fig. 10.1) obtained from a first injection, the temperature of the column being considerably greater than the ambient temperature. Its peaks are unresolved and appear a very short time after injection.

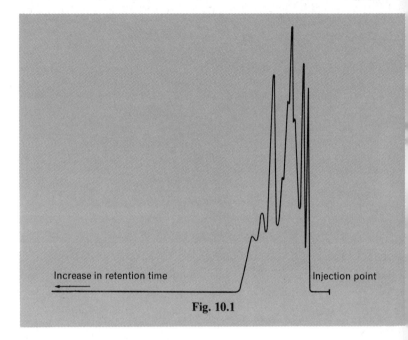

Increase in retention time   Injection point

Fig. 10.1

Question
Which one of the following factors would you change first of all to give a considerable improvement in the resolution?

1. reduce the carrier gas pressure ← page 232
2. reduce the column temperature ← page 237
3. change the stationary phase → page 249
4. increase the length of the column → page 256

You think that a decrease in the amount of stationary phase would rectify the thermal decomposition of the sample? Well, it would help to do so, as this course of action would result in shorter elution times for the components, thus reducing the contact time between the sample and the hot column. Unfortunately, some thermal decomposition of the sample could still occur in that short time.

What other action should be taken first to rectify thermal decomposition?

Return to → page 250, and choose this alternative.

You think another cause of tailing could be the use of the wrong stationary phase. Quite right. Strongly polar columns could cause this, or it could be caused by a strongly polar sample 'self-associating' by hydrogen bonding.*

Now look at the chromatogram (Fig. 10.6).

The peaks are too well separated, the retention time of each peak being longer than necessary, to give adequate resolution. If we could shorten these retention times, the chromatogram would be an ideal one for all purposes.

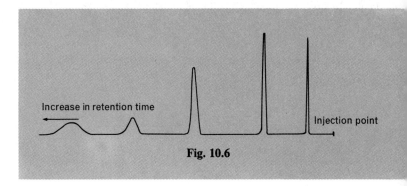

Increase in retention time

Injection point

Fig. 10.6

* Tailing is further discussed in Part Eleven, section 5. When a substance 'hydrogen bonds' to itself (i.e. **SELF ASSOCIATION**) the result is polymer like molecules with higher boiling points. If the substance can be made to 'hydrogen bond' to the stationary phase, 'tailing' is overcome.

Question

Although it would help to use a shorter column, changing a column is a time-consuming exercise. Would you say that the *quickest* way to shorten the retention times (and hence the total analysis time) would be:

1. increase the column temperature and increase the flow rate of the carrier gas ← page 234
2. increase the column temperature only → page 251
3. increase the flow rate of the carrier gas only? → page 259

You would choose none of these alternatives. Let's look again at the effects of temperature on an analysis.

We said that a change in temperature affects:
the partition coefficient,
the gas flow rate in the column for a given pressure ratio (P_i/P_o),
column efficiency.

Regarding column efficiency the rule is AS TEMPERATURE RISES, COLUMN EFFICIENCY DECREASES.

Do we need to increase or decrease the efficiency of the column to effect a separation between m-xylene (b.p. 139°C) and p-xylene (b.p. 138°C)?

Think about this then return to ← page 235 and choose one of the alternatives listed there.

No, a coarse packing would allow the carrier gas to pass more easily through the column. What effect would this have on the velocity gradient? Would it make it smaller or greater?

Think about this then return to ← page 229 and choose a better alternative.

No, a reduction in the column temperature would simply increase the analysis time and you were told that it is reasonable already. It is the selectivity of the chromatographic process which is at fault.

Which of the suggestions would most improve the chromatographic process?

If you do not know, then re-read ← page 74, then choose the correct alternative on ← page 237.

back ref. page 231

You think that too much sample has been injected. Wrong. If it had, what could you say about the shape of the peaks?

Is the resolution adequate in this chromatogram?

Answer these questions then return to ← page 231 and choose a better alternative.

A change in stationary phase may improve the resolution, but you should have taken into consideration affinity and polarity effects **before** you made up your column. A wrong choice of stationary phase would be revealed by a failure to separate the components of interest and a reappraisal of all the facts would help at this stage.

Since in the chromatogram shown, most of the peaks are distinctly visible, what can you say about the choice of stationary phase?

Can you remember what was said about the effect of a decrease in temperature on retention time? If you cannot, then read ← page 235 again.

Now return to ← page 242 and choose from the suggestions listed, the course of action which will give the MOST improvement in the resolution.

back ref. page 257

Correct. A reduction in the amount injected would reduce the area of the peaks and thus the resolution would be improved.

Sometimes the peaks in the first chromatogram appear asymmetrical as those in Fig. 10.4, or the asymmetry could be such that the rear profile of the peak is steeper than the front profile. The latter asymmetry occurs when using a column containing too few plates.

Other causes of asymmetry can be:

(i) an overloaded column, which can be corrected by injecting less sample, and

(ii) thermal decomposition of the sample on the column.

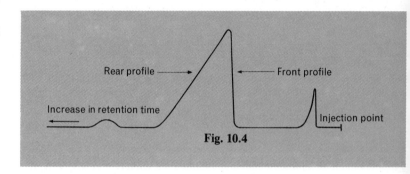

Fig. 10.4

Question
Would you say that thermal decomposition of the sample can be rectified by:

1. lowering the oven temperature? ← page 231
2. increasing the volume flow rate of the carrier gas? → page 260
3. decreasing the amount of stationary phase? ← page 243
4. none of these? ← page 239

Your answer – increase the column temperature – would certainly help to reduce the analysis time, but if it were coupled with an increase in the carrier gas flow this would be the *quickest* way of shortening the analysis time.

Return to ← page 244 and choose this dual method.

Your answer, to increase the carrier gas flow rate, would be no help here. Remember we said base line drift is more pronounced when the volatility of the stationary phase is high.

Would increasing the carrier gas flow rate suppress the volatility of the stationary phase?

Return to → page 263 and choose another alternative.

Yes, increasing the column temperature and carrier gas flow would shorten the retention time of peak B. This action would almost certainly increase the relative height of the peak too. (Peak A might go off the chart but do not worry about that, for you can measure its retention time and area on your first chromatogram; a better course of action would be to increase the attenuation at the new temperature until it is on the chart and then calculate the ratio of attenuation factors and use this in comparing B with A).

The chromatogram could contain a selection of some of these difficulties. Look at this one (Fig. 10.8). Clearly the analysis time is too long, but the early peaks are unresolved.

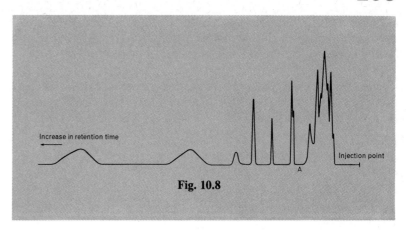

Increase in retention time

Injection point

A

Fig. 10.8

Question

To analyze this mixture would you say that we needed to:

1. separate the mixture into two or three fractions by prior distillation and chromatograph these separately? → page 268

2. use a lower temperature to resolve the early peaks, then commence to increase the temperature gradually from A onwards as quickly as is consistent with the maintenance of the full resolution of the later peaks? → page 263

3. use a lower temperature to resolve the early peaks and then increase the carrier gas flow to a higher level from A onwards, whilst maintaining the resolution of the later peaks? → page 266

You would say that the analysis time is adequate but the peak from component B is too small for useful interpretation. I would agree with you on the second point, but not on the first!

Since there are only two peaks in the chromatogram and their separation is more than adequate, the analysis time could be reduced until their **peak resolution** was just greater than unity.

Return to ← page 234 and make the correct choice.

You would change the stationary phase. Since the two compounds are isomers, affinity and polarity effects would influence both to a similar extent, therefore the choice of selective phases is limited.

Remember, we said that the rule was AS TEMPERATURE RISES, COLUMN EFFICIENCY DECREASES.

Do we need to increase or decrease the efficiency of the column to effect a separation between *m*-xylene (b.p. 139°C) and *p*-xylene (b.p. 138°C)?

Think about this, then return to ← page 235 and choose another course of action.

Your choice, increase the length of the column, would give an improvement in the resolution. The analysis time would be lengthened, but this wouldn't matter here as the analysis time is quite short anyway.

But g.l.c. is designed to give accurate results, *quickly*. Think of the inconvenience of making up a longer column, fitting it in the oven, etc.

There is a much more convenient way of producing a considerable improvement in the resolution.

Can you remember the relationship between temperature and retention time? If not, read ← page 235 again.

Now return to ← page 242 and choose a better alternative.

Yes, the poor resolution in this case, since the analysis time is reasonable, is caused by the wrong stationary phase. One would take into account chemical affinity and polarity effects in the choice of column.

Now look at the chromatogram (Fig. 10.3).

The resolution is still not good enough to give complete separation of the two major peaks. An improvement in the efficiency of the column would help. This could be achieved by lowering the oven temperature and by finding a new gas flow rate by trial. A longer column would improve things still further.

Unfortunately these steps will prolong the analysis time and it is already quite long.

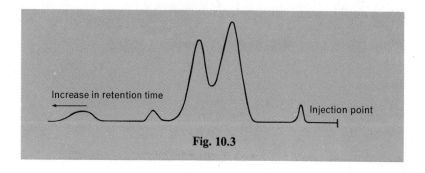

Increase in retention time

Injection point

Fig. 10.3

Question
Since we know that the area of each peak represents the quantity of that particular component in the mixture, would you say that an alternative to these suggestions would be

1. use less sample for the injection ← page 250
2. use more sample for the injection ← page 238
3. use a column containing less stationary phase → page 262

If you would do none of these, turn to → page 265

You would increase the column temperature. This would help for it would increase the height of peak B, although peak A might go off scale. Unfortunately, the analysis time is rather long too.

What could you do to shorten it?

Return to → page 269 and choose a better course of action.

Your answer – increase the carrier gas flow only – would certainly help to reduce the analysis time, but if it were coupled with an increase in the oven temperature, this would be the *quickest* way of shortening the analysis time.

Return to ← page 244 and choose this alternative.

Your answer, increase the volume flow rate of carrier gas, would *help* to rectify the asymmetry, as this action would result in shorter elution times for the components. This means that they would be in contact with the hot column for a shorter time, but thermal decomposition could still occur.

What other action should be taken first, to rectify thermal decomposition?

Return to ← page 250 and choose this alternative.

You would say that the analysis time is too long, but the peak from component B is large enough to be interpreted. I would agree with you on the first part, but not on the second! Look at Fig. 10.7A again.

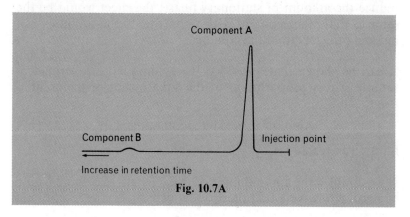

Fig. 10.7A

Peak B is small and rounded, so it would be difficult to measure its area, height or retention time accurately. In other words it is too small for useful interpretation.

Return to ← page 234 and make the correct choice.

You would use a column containing less stationary phase? This would not help the resolution.

At the moment the resolution is not good enough. If we were to reduce the amount of stationary phase, the effect would be the same as that of injecting MORE sample on to the column, or of reducing the length of the column.

It can be shown that, provided the column is not overloaded, sample size is proportional to the square root of the column length, i.e.:

$$\text{Amount injected } \alpha \sqrt{\text{column length}}.$$

for a given resolution.

If we doubled the amount injected, what length of column would be needed to ensure that there was no change in the resolution?

Consider your answer to this question, then choose a better way of improving the resolution from the suggestions listed on ← page 257.

back ref. page 253

Your choice, to use a lower temperature to resolve the early peaks and then to commence to increase the temperature gradually at a point (say A) whilst maintaining the resolution on later peaks, is quite right. Good, This is *temperature programming*.

Unfortunately, when adopting this procedure it is nearly always true that as the temperature rises the baseline will drift upwards (see Fig. 10.9).

This will be more pronounced the higher the volatility of the stationary phase.

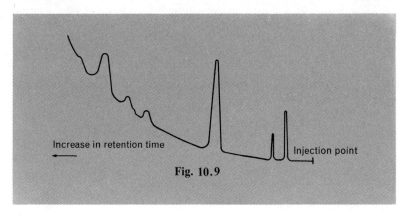

Fig. 10.9

Increase in retention time

Injection point

Question

To minimise baseline drift from this cause would you:

1. ensure that the upper temperature limit of the programmed run was less than the safe maximum operating temperature of the stationary phase? → page 267
2. increase the carrier gas flow rate? ← page 252
3. keep the upper temperature limit of the programmed run less than the safe maximum operating temperature of the stationary phase and use a phase with a higher boiling point and of the same polarity as that which gave the drift? → page 270

You would increase the flow of carrier gas? Since the analysis time is rather long, this would certainly shorten it, but would this action alone increase the size of peak B?

What additional factor would you have to alter to increase the size of peak B?

Return to → page 269 and choose a better alternative.

You would do none of these? Well I assure you a solution to the problem is listed there, but if you have a better idea discuss it with your supervisor.

You were told that the resolution was not good enough. Let us consider how to improve it further.

What is the relationship between the area of a peak and the quantity of sample injected?

If the quantity of sample injected is decreased, what happens to the peak?

Which would improve the resolution in this chromatogram, an increase or a decrease in the area of the peaks?

Is there an alternative listed on ← page 257 which if followed would fit in with your answers to the questions above? If there is, choose it.

back ref. page 253

You would firstly use a lower temperature to resolve the early peaks? Fine, this action would certainly improve the resolution of these peaks. Then, secondly, you would increase the carrier gas flow to a new higher level at A, yet maintain the resolution on these later peaks? I don't agree with you in this case.

This latter course would certainly shorten the analysis time of the later peaks and is known as *flow programming*. This procedure is often used when the sample components are unstable or would decompose at a higher temperature.

Which factor affects the size of the peak assuming the quantity of sample injected is kept constant?

Return to ← page 253 and choose a better alternative.

Your answer, to keep the upper temperature limit of the programmed run less than the maximum safe operating temperature of the stationary phase, is, of course, essential in a temperature programmed run, but look again at the question. It was 'How would you minimize base line drift caused by the high volatility of the stationary phase?'.

The vapour pressure of a stationary phase with a high volatility would increase as the operating temperature increased. We can only lower the vapour pressure of our stationary phase by reducing the operating temperature. If it is undesirable to lower the operating temperature, what other course can we adopt to decrease base line drift?

Look at the other alternatives on ← page 263 and choose that which fits the answer to the question above.

back ref. page 253

Your choice would be to separate the mixture into two or three fractions by prior distillation and chromatograph these separately.

This action would not exploit the advantages of gas–liquid chromatography; in addition there is a possibility of thermal degradation or oxidation of the sample in the distillation process unless it is performed under a vacuum or inert atmosphere.

No distillation process is as efficient as gas–liquid chromatography for the number of theoretical plates in even the best distillation columns is always smaller than that of a chromatography column.

Look carefully again at the chromatogram, think about the problem then choose a better alternative from the others listed on ← page 253.

Quite right. The analysis time is too long, and the peak from component B is too small for useful interpretation.

One reason why this peak is too small could be because there is only a little of component B in the sample mixture. One way to increase the area of peak B would be to inject a larger sample of the mixture. The danger here is that we might inject so much that the column would be overloaded (Partition theory no longer applies when the column is overloaded, since the solution of components A and B in the stationary phase is not in equilibrium with the vapour.)

Of course, there are other ways of increasing the height of the peak from component B. One way would be to concentrate component B by distilling some of the lower boiling component A from the mixture before injection.

Question
Would you say that a quicker way, and incidentally, one which would also shorten the analysis time would be to:

1. increase the column temperature?   ← page 258
2. increase the flow of carrier gas?   ← page 264
3. increase both column temperature and carrier gas flow?   ← page 253

back ref. page 263

You have chosen correctly. This combination would minimise base line drift during this procedure, which, since the rise in temperature is uniform with time, is called '*Linear Temperature Programming*'*. There is a piece of equipment which will do this automatically. It holds the temperature steady for a timed period (the time it takes for those early peaks to be resolved) then it heats the oven at a steady rate (which can be chosen by the operator) to a pre-set upper temperature. In this way a sample of the type illustrated, with many low and high boiling components, can be analysed readily.

Another method of dealing with base line drift spoiling the chromatogram would be to use a dual column system.†

Finally, if you obtain no peaks at all, check that you have a supply of carrier gas passing first to the inlet and then through the column. If not, then check the system for gas leaks or block ages in the appropriate part of the gas system.

A good way of checking your column is to keep handy a very short column (about 0·5 metres long) containing Apiezon L as stationary phase. This short column is substituted for the 'suspect' column and if the rest of the system is functioning satisfactorily, peaks will appear soon after an injection, confirming that the original column is defective.

Part 11 begins on → page 271

---

* S. D. dal Nogare and C. E. Bennet, *Anal. Chem.* **30**, 1157 (1958).
† E. Murand and W. A. Noyes, *J. Am. Chem. Soc.* **81**, 6405 (1959).

Part Eleven
Conclusion

(In this part, and in all the sections which follow, the pages are arranged in numerical order, and can be read in the same way as a normal text book.)

SOME years ago Professor Silvanus P. Thompson in his book 'Calculus Made Easy', headed one of his chapters 'Pitfalls, Dodges and Triumphs'. Every subject has its share of these and gas–liquid chromatography is no exception. Many of these are beyond the scope of this brief introduction to the subject, but as the student of the subject progresses and practices the art of gas–liquid chromatography, he will find sophisticated applications in the technical literature to solve most difficulties. To help to bridge the gap between this simple teaching programme and the technical literature, the following brief review has been written.

(1) The Preparation of Column Packings for gas chromatography is one of the most important steps in the art. One method of preparation has been described in the text (Part Six, page 115). The efficient drying of the coated packing is essential to the attainment of a high column efficiency. A fluidized drying technique has recently been described[1] by which it is claimed that there is assurance of (i) uniform coating, (ii) less fragmentation, (iii) expulsion of the fine particles and light impurities and (iv) a saving in time.

(2) Capillary Columns. These columns were not mentioned in the text, but can be made in nylon, glass, stainless steel or copper tubing about 0·25 to 0·5 mm i.d. and up to 200 m long. An open tubular capillary column consists essentially of a thin uniform coating of the stationary phase on the inside walls of the tube. Various techniques are described for applying the coating.[2–6] They are particularly useful where imperfect separation has been obtained on a short packed column, but better separation is required in a shorter analysis time. As an example, 100 ft. of capillary column will be able to separate compounds at least as well as 50 ft. of packed column of the same liquid phase, do it much more quickly, and pack into a much smaller oven.

(3) Sample Injection. The only technique described in the text is that using a syringe and septum. Accuracy and reproducibility of injection are very important factors in gas–liquid chromatography. A poor injection technique will affect both of these, since sample can be lost by evaporation, boil out from the needle, or blow back past the plunger. These losses are minimized by a smooth, quick injection technique. Modifications to the syringe technique taught in this programme are the subject of a review article by Hamilton.[7] The student is advised to try these modifications and choose the technique which suits him best. Details of syringes currently available are given in Appendix D, page 296.

For some applications, automatic injection is preferred, (e.g. the gas stream from a reactor can be analyzed automatically at regular intervals to measure product composition, thus providing close control over the chemical process.[8, 9, 10]

It is also possible to analyze volatile solid samples by gas–liquid chromatography.[11]

(4) Component Identification. The problem of positive peak identification is important in gas chromatography. Retention times are important but limited in their value when dealing with unknown compounds, unless some other analytical technique is used to supply further evidence as to structure and molecular weight.

Usually this evidence is given by combining the details given by Mass Spectrometry (which gives the molecular weight of the parent ion and a break down pattern of the molecule), Nuclear

Magnetic Resonance (which indicates the types of group present and their position relative to one another in the molecule) and Infra Red Spectroscopy (which gives a 'finger-print' spectrum of the molecule, which if it can be compared to a known standard spectrum identifies the component). In fact, there are now several commercially available instruments which couple g.l.c. and mass spectrometry, so that as each component is eluted from the column its mass spectrum is determined immediately.

But for many laboratories some, if not all, these techniques are in the luxury class and they have to rely on conventional chemistry to make a derivative of the component and identify it by the examination of its physical properties. Examples are the detection of unsaturation by bromination and the preparation of derivatives of alcohols.

Methods have been described[12] of splitting the effluent from the detector into several streams and reacting each one with a different functional group reagent. The disadvantage of this method is that it only identifies groups and not full structures.

A technique has been devised for the analysis of organic compounds by catalytic hydrogenation and gas–liquid chromatography.[13–16] Hydrogen sweeping the compound over a heated catalyst, saturates multiple bonds, and causes fission at the point of attachment of atoms such as halogen, oxygen, sulphur and nitrogen atoms. The resulting products, which are mainly the parent hydrocarbon and/or the next lower homologue, are identified by their retention time from an appropriate, interconnected gas–liquid chromatography column. The method has been applied to solids and organometallic compounds, e.g. lead alkyls.[17, 18]

(5) **Derivatives.** Difficulties can be experienced with strongly polar molecules 'tailing' by the self association caused by hydrogen bonding. Acids and amines are notorious for this.

(i) *Acids*. Acids have been analyzed by using a glass column 1·5 m long, $\frac{1}{4}''$ o.d. packed with 5% 15 to 20,000 carbowax and 5% isophthalic acid on 60 to 100 mesh Embacel, with nitrogen as carrier gas.[19, 20]

The polarity of an acid can be reduced by preparing its methyl ester, which invariably is easier to chromatograph in practice. There are four methods in use, three of which are reviewed by Vorbeck.[21] The diazomethane method is a clear winner, but has the disadvantage of toxicity and inflammability for the beginner. A simpler method of preparing esters is that of Metcalf and Schmitz using BF_3– methanol reagent. The preparation is done in a test tube by adding the reagent to the acid, refluxing for 2 minutes and extracting the ester with solvent.[22] The fourth method involves direct injection of the tetramethyl ammonium salt of the acid into the column. The methyl ester is formed by pyrolysis of the salt in the heated injection port, or on the first plate of the column.[23]

A method used in ICI Petrochemical & Polymer Laboratory at Runcorn devised by D. J. Rees for dibasic acids (it can also be used for monobasic acids) involves the addition of a chloroform/methanol, nitric acid mixture, to the acid. The solution is refluxed for 12 minutes, water washed twice, dried, and the solvent is removed by warming gently in a stream of air. Conversions of 96 to 99% are reported with adipic, glutaric, and succinic acids.

Details of the quantitative chromatographic analysis of fatty and resin acid methyl esters have been reported recently.[24]

J. S. Parsons recommends the preparation of sulphonyl fluoride derivatives for the separation and identification of sulphonic acids by gas chromatography because of their better volatility and stability than the acids. The method is illustrated with naphthalene mono, di and tri sulphonyl fluoride and benzene trisulphonyl fluoride.[25]

(ii) *Amines*. Details of the separation of aliphatic amines have been described.[26,27,28] A. R. Oldham in ICI Petrochemical & Polymer Laboratory, Runcorn advises the use of the dodge of using low concentrations of stationary phase e.g. 2% in a packed column, to successfully chromatograph amines.

Nineteen industrially important aryl amines contained in a complex product mixture were identified by chromatographing their N-trifluoroacetyl derivatives in a single 70 minute run using a column prepared from a blend of Carbowax 20M and Apiezon L.[29]

(iii) *Amino Acids*. As amino acids are involatile they have to be converted to a volatile derivative for g.l.c. analysis. Pollock separated 14 amino acids using the N-trifluoroacetyl amino acid n-butyl esters on a capillary column containing Carbowax 20M.[30] Amino acids have also been analyzed recently by preparing their methyl esters and using dimethyl dodecanedioate as internal standard.[31]

(iv) *Ethanolamines*. These too have been analyzed quantitatively by gas–liquid chromatography as the trifluoroacetyl derivatives.[32]

(v) *Olefines*. Silver nitrate dissolved in ethylene glycol has been in use for years to give useful separation of olefines due to the loose π-complexes formed by the silver ions and each olefine. The temperature, however, is limited to 70°C, because above this the silver nitrate decomposes to silver metal and loses its activity.

One other way of dealing with double and multiple bonds is by catalytic hydrogenation (described in Section 4).[33]

These then are a few of the pitfalls and some of the dodges of gas–liquid chromatography. The triumphs are in your hands. To quote Professor Thompson again:

> "what one fool can do, another can."

T

References

1. R. F. Kruppa *et al.*, *Anal. Chem.* **39**, 851 (1967).
2. M. J. E' Golay, *Gas Chromatography*, I.S.A. Symposium, August 1957. (Eds. V. J. Coates, H. J. Noebels and I. S. Fagerson), Academic Press, New York, 1958. p. 1.
3. D. H. Desty, A. Goldup and B. H. F. Whyman, *J. Inst. Petrol* **45**, 287 (1959).
4. R. P. W. Scott, *Nature* **183**, 1753 (1959).
5. A. Zlatkis and J. E. Lovelock, *Anal. Chem.* **31**, 620 (1959).
6. D. W. Grant, *Gas Chromatography* 4th. Wilkens Symposium, May 1966 (E. Baumann, Sec.) p. 3.
7. C. Hamilton, *Laboratory Management*, Sept. 1965.
8. E. P. Sampsel and J. C. Aldrich, *Anal. Chem.* **31**, 1288 (1959).
9. F. A. Keidel and C. D. Lewis, *Anal. Chem.* **33**, 1456 (1961).
10. D. W. Grant, *Anal. Chem.* **36**, 1519 (1964).
11. C. J. Thompson *et al.*, *Anal. Chem.* **37**, 1042 (1965).
12. J. T. Walsh and C. Merritt, *Anal. Chem.* **32**, 1378 (1960).
13. M. Beroza and R. Sarmiento, *Anal. Chem.* **35**, 353 (1963).
14. M. Beroza, *Anal. Chem.* **34**, 1801 (1962).
15. M. Beroza and R. Sarmiento, *Anal. Chem.* **36**, 1745 (1964).
16. C. J. Thompson *et al.*, *Anal. Chem.* **37**, 1042 (1965).
17. N. L. Soulages, *Anal. Chem.* **38**, 28 (1966).
18. N. L. Soulages, *Anal. Chem.* **39**, 1340 (1967).
19. K. M. Fredricks, *A. van Leeuwenhock (J. Microbiol, Serol.)* **33**, 44 (1967).
20. J. R. P. Clark and K. M. Fredricks, *J. Gas Chromatog.* **5**, 99 (1967).
21. M. L. Vorbeck, *et al.*, *Anal. Chem.* **33**, 1512 (1961).
22. L. D. Metcalfe and A. A. Schmitz, *Anal. Chem.* **33**, 363 (1961).
23. D. T. Downing, *Anal. Chem.* **39**, 218 (1967).
24. F. H. M. Nestler and D. F. Zinkel, *Anal. Chem.* **39**, 1118 (1967).
25. J. S. Parsons, *Advances in Gas Chromatography*, (Ed. A. Zlatkis), Preston Technical Abstracts, Evanston, Illinois, 1967, pp. 88–90.
26. A. T. James and A. J. P. Martin, *Brit. Med. Bull.* **10**, 170 (1954).
27. A. T. James, *Biochem. J.* **52**, 242 (1952).
28. A. T. James, *Anal. Chem.* **28**, 1564 (1956).
29. R. A. Dove, *Anal. Chem.* **39**, 1188 (1967).
30. G. E. Pollock, *Anal. Chem.* **39**, 1194 (1967).
31. M. Gee, *Anal. Chem.* **39**, 1677 (1967).
32. L. E. Brydian and H. E. Persinger, *Anal. Chem.* **39**, 1318 (1967).
33. E. A. Walker, *Gas Chromatography* 4th. Wilkens Symposium, May 1966 (E. Baumann Sec.) p. 15.

Summary and Exercises

Answers to the questions in this section are on page 282.

Part 1 'Boiling Points and Vapour Pressure'

When the vapour pressure of a liquid equals that of the atmosphere, it boils. A mixture of liquids may be separated by condensing their vapours at their individual boiling points.

Separation is easy when the difference between the boiling points of the components of the mixture is great.

Separation is most difficult when the boiling points of the components of the mixture differ by only a small amount.

Gas–liquid chromatography depends partly on the fact that each liquid has its own particular boiling point.

Question

1. Would it be easy to separate *cis*-decalin (b.p. 195°C) and *trans*-decalin (b.p. 185·5°C) by distillation? If not, why not?

Part 2 'Partition Coefficients'

The Partition Coefficient K =

$$\frac{\text{concentration (i.e. g/ml) of solute dissolved in solvent A}}{\text{concentration (i.e. g/ml) of solute dissolved in solvent B}}$$

The numerical value of K depends on the solvent mixture used. The constituents of a solute mixture are called the COMPONENTS. The constituents of a solvent mixture are called PHASES, (assuming that the constituents are immiscible).

Questions

1. Calculate the partition coefficient K, of a solute S between water and benzene at 20°C from the following table of results, which gives the number of grams of solute S dissolved in 100 ml of each solvent.

Water	0·635	1·023	1·635	2·694
Benzene	0·550	0·898	1·450	2·325

2. Mercuric bromide (0·036 g) is completely dissolved in a mixture of 100 ml of water and 100 ml of benzene. If the partition coefficient for the system is 0·88 how much mercuric bromide would be dissolved in the benzene layer, given that mercuric bromide is more soluble in benzene than water?

3. How much bromine would be extracted from 50 ml of a saturated solution of bromine water at 20°C by 5 ml of carbon disulphide? (The solubility of bromine in water at 20°C is 0·0358 g/ml. The partition coefficient for the system is 76·4. Bromine is more soluble in carbon disulphide than water.)

4. 500 ml of water contain 100 g of succinic acid in solution. How many ether extractions of 250 ml each would be required to extract at least 50 g of the succinic acid from the aqueous solution, given that the partition coefficient of succinic acid between water and ether is 1/7, that the acid is more soluble in water, and the temperature stays constant.

Part 3 'Chromatography'

GAS–LIQUID CHROMATOGRAPHY is the process of separating the components of a mixture by making use of their partition coefficients between a gaseous moving phase and a liquid stationary phase.

The stationary phase is deposited on an inert support material. In the chromatographic process, a vapour A which is soluble in both phases, would be extracted from the liquid stationary phase by the gaseous moving phase. This can be represented graphically by a curve which is called an ELUTION CURVE (or PEAK) (Fig 4.2 ← page 58).

At point M (the maximum) a definite volume V of carrier gas, called the retention volume will have passed through the column, and appeared at the detector.

The RETENTION TIME =

$$\frac{\text{RETENTION VOLUME}}{\text{VOLUME FLOW RATE OF THE CARRIER GAS}}$$

and is represented by the distance OP on the X axis of the graph.

Some factors which affect the RETENTION TIME are summarized on ← page 51.

The EFFICIENCY of a column is measured by the width of the peak, the narrower the peak, the higher the efficiency and vice versa.

Part 4 'Detection'

A brief description of detectors in use will be found on ← pages 70 and 71. If n is the NUMBER OF THEORETICAL PLATES in a column, then

$$n = 16 \left(\frac{t}{W_b}\right)^2$$

where t = retention time of a component peak, and W_b is its width at the base between the inflection tangents.

The HEIGHT EQUIVALENT TO A THEORETICAL PLATE (H) is

$$\frac{\text{length of column}}{n}$$

where n is the number of theoretical plates.

The TOTAL QUANTITY of each component which has passed through the detector is represented by the AREA under its peak in the chromatogram.

The PEAK RESOLUTION, R, between two peaks in a chromatogram is given by the formula

$$R = \frac{2(\text{distance from one max. peak height to the other})}{\text{sum of the base widths of each peak}}$$

Questions

1. In an experiment to find the value of H for a particular 2 metre column, the following results were obtained after drawing the inflection tangents to the peak obtained:
(i) peak width at base 9 mm.
(ii) retention time 84 mm.
Calculate the value of H from this information.

2. The following results were obtained from peaks in the same chromatogram after drawing the inflection tangents:

Compound	Retention time mm	Peak width at base mm
o-xylene	26·5	7·9
m-xylene	21·25	6·25
p-xylene	19·75	6·0

If the Height equivalent to a theoretical plate was 1·11 mm, what was the length of the column in mm?

3. The Peak Resolution for two peaks A and B is 1·52. After drawing the inflection tangents, their peak widths at the base are found to be 5 mm and 3 mm respectively. The retention time of peak A is 11 mm. What is the retention time of peak B?

4. Calculate the Peak Resolution R for the two peaks C and D from the table of results below (assume that the peak widths given are those between the inflection tangents).

Peak	Retention Time mm	Peak width at base mm
C	19	5·5
D	24	8

Would you say these peaks were well resolved or not? If not, why not?

Part 5 'The Stationary Phase'

The STATIONARY PHASE should be:
(i)   a chemically stable liquid over a wide range of temperature,
(ii)   of a low vapour pressure over a wide range of temperature,
(iii) inert with respect to the solutes, and
(iv)   a good solvent for the components.
POLARITY: polar substances will dissolve in other polar substances; non-polar substances will dissolve in other non-polar substances; BUT a polar substance will NOT normally dissolve in a non-polar one, and vice versa.
HYDROGEN BONDING: this occurs when a hydrogen atom can form a bridge between two ELECTRONEGATIVE atoms such as fluorine, oxygen and nitrogen. Parts of the *same* molecule, or two different molecules can be linked up by these bonds.

Taking into account the effects of POLARITY and HYDROGEN BONDING, it can be said that *by careful choice of stationary phase a difference in activity can assist a difference in vapour pressure for the separation of mixtures of solutes by gas–liquid chromatography. A table of group 'polarity' is given on* ← page 97.

In general, we can say LIKE DISSOLVES LIKE

Part 6 'The Preparation and Packing of Columns'
THE STATIONARY PHASE is coated on to a support material which should be inert, porous, of uniform particle size within the limits 60 to 100 mesh and have a specific area of 2 to 5 metres^2/g. Examples are crushed firebrick,* kieselguhr, Celite, etc.

For normal use the concentration of stationary phase on the support material is 5 to 30 $w/w\%$ (Concentrations below 5% can be used, however, for certain applications.)†

Uniform application of stationary phase to the support material is ensured by making a wet slurry of it in a suitable volatile solvent, e.g. pentane, ether or acetone. The solvent is removed by shaking and warming the slurry under vacuum.

The COLUMN is made by packing the coated support material uniformly into a tube made of glass, plastic or metal (e.g. copper, aluminium, stainless steel).‡ Glass columns are pre-coiled and filled by gently vibrating them whilst under slight vacuum. Other columns are filled in a straight length or bent into a 'U' shape and, if necessary, coiled afterwards on a former.
Column dimensions can vary widely according to the use intended but a useful size is $\frac{3}{16}$ in. o.d. × 1 to 2 metres in length. The column must be conditioned before use.

---

* At low concentrations of stationary phase, crushed firebrick is not recommended, since it has certain catalytic properties. At high concentrations, however, these become innocuous.   † ← page 272.   ‡ Nylon is used too, mainly for capillary columns (see Conclusion, ← page 271).

Part 7 'Sample Injection and Syringe Technique'*

SYRINGE TECHNIQUE

A precision 10 microlitre syringe is used to inject the sample into the gas chromatograph. When using it the following precautions should be observed:

(i)   Do not touch the stem of the plunger. Hold it only by the top.
(ii) Do not bend the plunger.
(iii) Do not bend the needle.
(iv) Wash out the barrel with solvent after use.
(v)   Store the syringe in its case, away from dust and contamination.

SAMPLE INJECTION

(i)   Insert the needle into the sample and draw some of it into the barrel.
(ii) Remove the air bubbles, and draw in a few microlitres.
(iii) Wipe the outside of the needle.
(iv) Insert the needle all the way into the rubber septum, and depress the plunger rapidly, at the same time mark the injection point on the chart (chromatogram).

Part 8. 'Qualitative Analysis'

A simple 'block' diagram of the complete gas–liquid chromatograph is given on ← **page 155.**

The retention time of the AIR PEAK is the smallest possible, since it does not dissolve in the stationary phase, but passes straight through the column in the gaseous moving phase. The retention times of other components can be measured relative to that of the air peak. These are called RELATIVE RETENTION TIMES.

* See also page 271 and Appendix D page 296.

The EFFECTIVE NUMBER OF THEORETICAL PLATES, N in a column is given by the formula:

$$N = n \left(\frac{\text{retention time of a component relative to the air peak}}{\text{retention time of that component}} \right)^2$$

where n is the number of theoretical plates in the column.

A guide to the amount of stationary phase to use, based on the value of x, the retention time of the air peak, and y the retention time of a component relative to that of the air peak, found by using a column containing 10 $w/w\%$ of stationary phase, is as follows:

If $y > 7x$, use less stationary phase (down to about 5 $w/w\%$)
If $y \rightarrow x$, use more stationary phase (up to about 30 $w/w\%$)
If $x < y < 7x$, then the quantity of stationary phase used, (i.e. 10 $w/w\%$) is probably adequate, *provided the temperature of the analysis is not changed.*

In **general** it can be said:

(i)   When the BOILING POINTS of the solute compounds are HIGH (i.e. LOW VOLATILITY) use a SHORT COLUMN, and vice versa.
(ii) The FIRST CHOICE OF OVEN TEMPERATURE is the MAXIMUM SAFE TEMPERATURE FOR THE STATIONARY PHASE IN USE.
(iii) Although the AMOUNT OF ANY COMPONENT present is represented by the AREA UNDER ITS PEAK IN THE CHROMATOGRAM, we can APPROXIMATE THE AREA in all but the most accurate work TO PEAK HEIGHT.

The Procedure for Qualitative Analysis
(i) Obtain a chromatogram at $x°C$ and v cm³/min containing n peaks.
(ii) Measure the retention time of each peak, $t_1, t_2 \ldots t_n$ and the maximum peak height, $h_1, h_2 \ldots h_n$.
(iii) Select the material thought to be present in the mixture, and add a little of it to a portion of the mixture.
(iv) Obtain a new chromatogram of the enriched mixture. If it contains
  (a) $(n + 1)$ peaks, then the selected material was NOT in the original mixture.
  (b) n peaks, ONE OF WHICH HAS INCREASED IN PEAK HEIGHT RELATIVE TO THE OTHERS, then that selected material is MOST PROBABLY IN THE ORIGINAL MIXTURE, *provided* the RETENTION TIMES, $t_1, t_2, \ldots t_n$ are the same, and the conditions i.e. $x°C$ and v cm³/min are kept constant.
  (Complete identification is provided by analyzing that component by Mass Spectrometry, Nuclear Magnetic Resonance and Infra Red Spectroscopic techniques).

A set of INTERNAL STANDARDS has been set up by the IUPAC Committee. These are quoted on ← **page 195**. Retention times are often quoted relative to these standards. To simplify the system, Kovats proposed a retention index based on the normal alkanes as reference substances. By definition the Kovats Index of the normal alkanes is 100 times the number of carbon atoms present in the molecule. For other substances, KOVATS INDEX, I, is given by the formula:

$$I = 100 \left[n \left(\frac{\log R_x - \log R_z}{\log R_{z+n} - \log R_z} \right) + z \right]$$

where R_x is the retention time of an unknown substance X
  R_z is the retention time of the normal alkane having z carbon atoms
  R_{z+n} is the retention time of the normal alkane having $z + n$ carbon atoms
and n is the difference in the number of carbon atoms for the normal alkanes.

Kovats Index can also be obtained by the use of a nomogram (→ **page 289**).

Questions
1. Calculate the effective number of theoretical plates in a particular column, given that the retention time of the air peak is 2 mm and that of a single substance 81 mm, and the peak width at the base (obtained by drawing the inflection tangents) is 9 mm.
2. Calculate the length of the column on which the following data were obtained: Retention time of the air peak, 0·5 mm, retention time of a substance 20 mm; number of effective theoretical plates 262; and height equivalent to a theoretical plate 0·98 mm.
3. The following values were obtained from a chromatogram:

Peak heights (mm)			Retention times (mm)		
Peak A	Peak B	Peak C	Peak A	Peak B	Peak C
115	108	140	9·5	11·5	26·5

Under the same conditions the retention times of *n*-butyl bromide and toluene are 9·5 and 11·5 mm respectively.

A little toluene was added to the mixture and a new chromatogram obtained, containing three peaks, one of which was 120 mm in height whilst the other two peaks were about the same height as before. Which component was identified as possibly toluene, A, B or C?

4. In an analysis of a sample taken from a bottle marked decalin, two peaks A and B, were obtained whose retention times were 52 and 57·5 mm respectively. Under the same conditions the retention time of decane used as standard was 25·5 mm and the retention time of the air peak was 1·5 mm. Given that the retention time of *trans*-decalin relative to decane is 2·1, which component was identified as possibly *trans*-decalin?

5. Calculate Kovats Index, I for substance S whose retention time is 74 mm, given that on the same column, under the same conditions, the retention times of octane and decane are 45 and 111 mm respectively.

6. The table lists the Kovats indices for the stationary phase Apiezon L at 160°C.

Two component mixture		IUPAC International (Internal) Standards		
Propyl Butyrate b.p. 143°C	Butyl Butyrate b.p. 166·6°C	*p*-xylene b.p. 138°C	Cyclo-hexanone b.p. 155·65°C	Cyclo-hexanol b.p. 161·1°C
850	951	924	913	900

Which standard would be the best choice for measuring the relative retention times of the two component mixture?

7. Using hexane and octane, whose retention times are 96 and 270 seconds respectively, calculate the Kovats Index from the nomogram on → **page 291** for the substances listed, the stationary phase and conditions being constant.

Substance	Retention Time (sec)
(a) Propionaldehyde	252
(b) Methyl propionate	420
(c) Methyl Ether	66
(d) Methylal	198

Part 9 'Quantitative Analysis'
Criteria for INTERNAL STANDARDS *for quantitative analysis*
The standard must:
(i) be miscible with the sample mixture being analyzed
(ii) it must not react with any component
(iii) it should give only one peak, which should not overlap any sample component peak
(iv) have a retention time close to the sample component of interest.

For the most accurate work the PEAK AREA represents the total amount of each component that has passed through the detector. Peak areas can be measured in several ways. If u and s signify 'unknown' and 'standard' peaks respectively, providing the peaks are Gaussian as in Fig 9.2A ← **page 215**.
(a) measure the peak heights, h and the peak width at half peak height w. Then the Peak Area Ratio

$$\frac{A_u}{A_s} = \frac{h_u w_u}{h_s w_s}$$

(b) Draw the inflection tangents LM and MN for each peak. Measure the heights PM $= l$, and the width at the base LN $= v$ for each peak. Then the Peak Area Ratio

$$\frac{A_u}{A_s} = \frac{l_u v_u}{l_s v_s}$$

(c) The areas can be measured by a planimeter or an integrator.
(d) The areas can be cut out with scissors and weighed accurately.

Since the PEAK HEIGHT, h, is proportional to the PEAK AREA provided the column is not overloaded, it is quite often used instead, and in that case the PEAK HEIGHT RATIO is given by h_u/h_s.

A CALIBRATION GRAPH on which the peak area or peak height ratio is plotted against $w/v\%$ pure component is used to determine the quantity of the unknown substance present. The graph is linear between about 1 to $30w/v\%$ of sample.

Questions

1. With four ether extractions of 250 ml each, a product P was extracted from a reaction mixture. A solution of P (1·14 g) in ether (50 ml) was prepared. Samples of 1 μl from each solution were analyzed by g.l.c. The peaks obtained were cut out from the chart paper and weighed. The results were:

Weight of peak from the solution of unknown strength =

0·0090 g

Weight of peak from the solution of known strength =

0·0106 g

Calculate the amount of P extracted from the reaction mixture.

2. Ether (660 ml) was used to extract a product P from a reaction mixture. Four standard solutions each containing 1, 2, 3 and 5 $w/v\%$ of P together with 0·5 ml of an internal standard S, were made up to 10 ml each with ether. A fifth solution was made up containing the same volume of standard S made up to 10 ml with the ether extract containing product P, from the reaction mixture. Each solution was analyzed by g.l.c. under identical conditions. The inflection tangents were drawn for each peak obtained, and the results are shown in the table

$w/v\%$	l_u mm	v_u mm	l_s mm	v_s mm	$\frac{A_u}{A_s}$
1	3·3	4·25	17·25	3·25	
2	6·15	4·25	16·85	3·50	
3	7·1	4·25	12·9	3·25	
5	13·2	4·10	15·4	3·25	
unknown	12·3	4·25	17·8	2·75	

Calculate A_u/A_s, the peak area ratio, and plot it against $w/v\%$ of P. From this calibration graph and the information given above, what volume of P was extracted from the reaction mixture?

3. A substance U was extracted from a reaction mixture with a total volume of 800 ml of ether. Four standard solutions each containing a different amount of U and a fixed amount of internal standard S were made up (solutions numbered 1 to 4 in the Table)

in 10 ml of ether. Solution 5 consisted of 5 ml of the ether extract together with the same fixed amount of internal standard S, made up to 10 ml with pure ether. On chromatographing all five solutions under the same conditions the following results were obtained (inflection tangents were drawn for each peak):

Soln. No.	h_u mm	h_s mm	l_u mm	v_u mm	l_s mm	v_s mm	$\dfrac{h_u}{h_s}$	$\dfrac{A_u}{A_s}$	wt of U in g
1	46·0	63·0	50·5	3·7	68·5	6·0			0·160
2	19·0	66·0	21·5	3·5	75·5	5·5			0·064
3	9·0	69·5	11·0	3·5	76·5	6·0			0·032
4	36·0	90·5	41·5	3·5	96·5	5·5			0·096
5	23·0	87·5	26·5	4·0	93·0	6·3			

Calculate the Peak Height Ratio h_u/h_s and the Peak Area Ratio A_u/A_s for each solution. Draw two calibration graphs:
(i) h_u/h_s against $w/v\%$ of U, and
(ii) A_u/A_s against $w/v\%$ of U.
From these deduce the weight of U extracted from the reaction mixture. What is the difference between the amounts obtained from each calculation?

Part 10. 'Interpretation of the Chromatogram'

A PRESSURE GRADIENT (and because of this a VELOCITY GRADIENT) exists in a column, which affects RETENTION TIMES. If P_i and P_o are the inlet and outlet pressures respectively, then ideally

$$\frac{P_i}{P_o} \simeq 1$$

This will be achieved more easily if the column works under increased pressure and not under vacuum.
The factors which affect RETENTION TIME are:

No.	Factor	Retention Time Short	Retention Time Long
1	Volume flow rate of the carrier gas	high	low
2	Concentration of stationary phase on the support material	low	high
3	Solubility of component in the stationary phase	low	high
4	Volatility	high	low
5	Temperature (oven)	high	low
6	Length of column	short	long

Details of individual chromatograms are given in the text commencing on ← page 242.

Answers.

Part 1.

1. No. Because at the boiling point of *trans*-decalin (185·5°C) the vapour pressure of the *cis*-decalin (b.p. 195°C) would be almost 760 mm, and this would make the separation by distillation difficult.

Part 2

1. 1·145. 2. 0·019 g 3. 1·58 g 4. 10.

Part 4.

1. 1·43 mm 2. 200 mm 3. 17·08 mm 4. 0·74; Not well resolved; Because the Peak Resolution is less than 1.

Part 8.

1. 1360 2. 250 mm 3. B 4. A 5. 910 6. Cyclohexanol
7. (a) 784; (b) 886; (c) 524; (d) 738.

Part 9.

1. 19·4 g 2. 32 ml 3. 1·64 g

Check your answers to the questions on ← **page 52** against those below.

1. Differential and Integral detectors.
2. The signal is proportional to the mass of the component which passes through the detector per unit time.
3. The katharometer consists of a thin wire which is part of a Wheatstone bridge circuit.
4. The wire is heated electrically. Variations in the thermal conductivity of the surrounding gas cause variations in the equilibrium temperature of the wire and consequently also in its resistance.
5. The flame detector.
6. The density of the gas leaving the column (column effluent) is continually compared to that of pure carrier gas.
7. The flame ionization detector.
8. They are sensitive, reliable and accurate.

If you have scored 6 or more correct answers, then turn to ← page 58. If less than 6, re-read the narrative on ← pages 70 and 71 and try the questions on ← page 52 again.

Check your answers to the questions on ← **page 141** against those below.

1. The supporting material for the stationary phase should be inert, porous, of uniform particle size, and with a specific area of 2 to 5 metres^2/g. (3 *marks*)
2. Uniform particle size.
3. 5 to 30 parts by weight of stationary phase to 100 parts by weight of the support material.
4. To ensure that the supporting material is uniformly coated with the stationary phase, the phase is first dissolved in a sufficient volume of a volatile solvent to cover the support material entirely. This solvent is then evaporated by careful warming and shaking under vacuum.
5. $\frac{3}{16}$ in. o.d. copper tubing is normally used.
6. To ensure uniform packing.
7. The packing density should be about 3 g/metre.
8. It must be 'conditioned'. (1 *mark*) This means that any excess solvent or volatile materials from the stationary phase must be removed by connecting the column to the chromatographic apparatus, and passing carrier gas through it for about 12 hours, whilst the column is held at a temperature about 10°C below the maximum temperature recommended for that particular stationary phase. (2 *marks*)

Credit each correct answer with 1 mark except where indicated above to the contrary. This gives a total of 12 marks for a complete list of correct answers.

If you have scored 10 or more correct answers then turn to ← page 143. If less than 10, then re-read the sequence on ← pages 115 to 140 and try the questions on ← page 141 again.

284

Check your answers to the questions on ← page 153 against those below.

1. A precision 10 microlitre syringe.
2. A few microlitres or less.
3. (a) All air bubbles must be removed from the barrel of the syringe. ($\frac{1}{2}$ *mark*)
   (b) The outside of the needle must be wiped with absorbent paper ($\frac{1}{2}$ *mark*).
4. The needle is inserted all the way into the rubber septum of the injection port, and emptied by rapidly pressing the plunger. After the needle has been emptied, the injection point is marked on the chromatogram at the same time as the needle is withdrawn from the rubber septum. (3 *marks*)
5. di-ethyl ether.
6. The excess solvent is removed by inserting the needle into a rubber serum cap which is attached to a line under vacuum.
7. (a) Avoid bending the needle or plunger at all times.
   (b) Do not touch any part of the plunger except the top.
   (c) Store the syringe in its case, away from dust and/or chemical contamination. (3 *marks*)
8. The proof of a successful injection is that a chromatogram with suitably shaped peaks will be obtained.

Credit each correct answer with 1 mark except where indicated above to the contrary. This gives a total of 12 marks for a complete list of correct answers. If you have scored 10 or more correct answers, then turn to ← page 155*. If less than 10, then re-read the sequence on ← pages 143 to 152, and try the questions on ← page 153 again.

---

* Details of the types of syringes currently available are given in Appendix D on page 296.

Check your answers to the questions on ← page 166 against those below.

1. (a) regulated carrier gas supply
   (b) pressure gauge
   (c) injection port
   (d) rubber septum
   (e) oven
   (f) column
   (g) detector
   (h) electrical signal
   (i) recorder
   (j) pen.   (10 *marks*)
2. The regulated pressure of the carrier gas supply is 5 to 10 psi.
                                                                     (1 *mark*)
3. (i) The sample is injected into the injection port where it mixes with the carrier gas and is carried over into the column located in the oven.   (2 *marks*)
   (ii) on the column, chromatographic separation takes place.
                                                                     (1 *mark*)
   (iii) the detector responds to each component as it is eluted by giving an electrical signal.   (1 *mark*)
   (iv) the recorder displays this signal as a visual trace.
                                                                     (1 *mark*)
4. (i) carrier gas.   (1 *mark*)
   (ii) electricity.   (1 *mark*)

Credit each correct answer with the appropriate mark. This gives 18 marks for a complete list of correct answers. If your score is 16 or more correct answers, then turn to ← page 156. If less than 16, then re-read the information on ← page 155 and try the questions on ← page 166 again.

Check your answers to the Criterion Test against those given below:

	Individual Marks	Total Marks

1. Any seven terms discussed in the programme e.g. partition, stationary phase, support material, peak height, retention time, Kovats Index, etc. — 3

2. When the components of a mixture are separated by making use of their differences in partition coefficients between two phases, the term CHROMATOGRAPHY is applied. — 2

3. (a) A chemically stable liquid over a wide range of temperature.
   (b) Inert with respect to the solutes.
   (c) Low vapour pressure.
   (d) A good solvent for the components. — 4

4. Carbowax 20,000. — 2

5. (a) The support material should be inert, porous, of uniform particle size.
   (b) A specific area of 2 to 5 metres^2/g.
   (c) Use 5 to 30 parts by wt. of stationary phase to 100 parts by wt. of supporting material.
   (d) Ensure supporting material is uniformly coated with stationary phase.
   (e) Ensure uniform packing of the column.

   Any 5, 1 mark each

   (f) A packing density of about 3 g/metre.
   (g) Column must be conditioned before use. — 5

6. (a) Insert needle into sample and draw sample into the barrel.
   (b) Remove air bubbles, and draw in a few μl when all air expelled.
   (c) Wipe outside of needle.
   (d) Insert needle into rubber septum all the way.
   (e) Depress plunger rapidly.
   (f) Instantaneously mark injection point on the recorder chart and withdraw needle.

   $\frac{1}{2}$ mark each — 3

7. A chromatogram with suitably shaped peaks will be obtained. — 1

8. (a) Do not touch the stem of the plunger, only touch the top of it.
   (b) Do not bend the plunger.
   (c) Do not bend the needle.
   (d) Store it in its case, away from dust or contamination.
   (e) Wash out the barrel with a suitable solvent after use.

   Any 3, 1 mark each — 3

9. (i) $n = 16(t/W_b)^2$ where t = retention time and W_b = peak width.

	Individual Marks	Total Marks

(ii)
$$R = \frac{2(\text{distance from one max. peak ht. to the other})}{(\text{Sum of base width of each peak})}$$
$$= \frac{2(PW)}{RS + YZ}$$

(iii)
$$K = \frac{\text{No. of g of solute dissolved in 1 ml of Solvent A}}{\text{No. of g of solute dissolved in 1 ml of Solvent B}}$$

or

$$\frac{\text{Concn. of solute in g/ml in Solvent } A}{\text{Concn. of solute in g/ml in Solvent } B}$$

(iv) Peak Area Ratio = $(A_u/A_s) = (h_u w_u / h_s w_s)$

where h_u = peak height (max.) of unknown

h_s = peak height (max.) of standard

and w_u and w_s are *peak widths at half peak height*

or

$$\frac{A_u}{A_s} = \frac{l_u v_u}{l_s v_s}$$

where l_u and l_s are heights of $\triangle$le formed by drawing inflection tang-

Any 3, 3 marks each

---

ents and v_u and v_s are width at base of that triangle.

(v) Retention time =
$$\frac{\text{Retention volume}}{\text{volume flow rate of carrier gas}}$$

Total Marks: 9

10. (a) Regulated carrier gas supply
(b) Pressure gauge
(c) Injection port
(d) Rubber septum
(e) Oven
(f) Column
(g) Detector
(h) Electrical signal
(i) Recorder
(j) Pen

Individual Marks: $\frac{1}{2}$ *each mark*

Total Marks: 5

11. (a) Sample injected into the injection port. — 1
(b) It is vapourized and mixes with the carrier gas.
(c) Sample is carried over into the column located in the oven. — 1
(d) On the column, chromatographic separation takes place. — 1
(e) The detector responds to each component as it is eluted by giving an electrical signal. — 1
(f) The recorder displays this signal as a visual trace. — 1

Total Marks: 6

12. The peak width. — 2

13. Differential Integral — 4

	Individual Marks	Total Marks
14.(i) (a) Thermal conductivity cell or Katharometer and,	1	
(b) Flame ionisation detector.	1	
(ii) *Katharometer* consists of a thin wire which is part of a Wheatstone bridge circuit. The wire is heated electrically. As the gas mixture passes over the wire there are variations in its thermal conductivity which cause variations in the equilibrium temperature of the wire and consequently its resistance. These changes are used to give a chromatogram.	3	
Flame ionisation detector. Combustion of the solute components in the carrier gas in a hydrogen flame causes ionization. The conducting gases cause a current to flow between two electrodes which are held at constant potential. This current is used to provide the chromatogram.	3	8

15.(i) *Polarity.* Since all polar substances dissolve in other polar substances and vice versa, a polar stationary phase will 'hold back' a polar substance in a mixture containing polar and non-polar substances, thus helping to separate substances more easily,

	Individual Marks	Total Marks
particularly if their boiling points and hence their vapour pressures are similar.	5	
(ii) *Hydrogen Bonding.* If one component of a mixture could form hydrogen bonds with the stationary phase, and the other(s) could not, this would help the separation by holding back that component thus allowing the separation to take place more easily, particularly if the boiling points of the constituents of the mixture are close together. Also, if, say a two component mixture contained two polar substances one of which could form hydrogen bonds with the stationary phase, this again will help the separation.	5	
Summarising these effects by careful choice of stationary phase a difference in activity can assist a difference in vapour pressure for the separation of mixtures of any solutes.	1	11
16.(i) (a) Efficiency would be increased.	1	
(b) Retention time would be longer.	1	
(ii) The maximum safe working temperature of the stationary phase.		
(iii) An increased carrier gas flow would	2	

	Individual Marks	Total Marks

ensure the components would be eluted more quickly. — 1

A decrease in carrier gas flow would ensure the components would be eluted more slowly. — 1 — 6

17. (a) Obtain a chromatogram, at a certain temperature and volume flow rate of carrier gas, containing n peaks. — 1

(b) Measure the retention times of the peaks, and note the max. peak heights. — 1

(c) Select substance which is thought to be present in the mixture, and add a little of it to the mixture. — 1

(d) Obtain a new chromatogram of the enriched mixture. If it contains $(n+1)$ peaks then the selected substance was not in the original mixture. — 1

(e) If it contains n peaks, one of which has increased in peak height relative to the others then that substance is most probably in the original mixture, if the retention times are the same, assuming the conditions do not change. — 2 — 6

18. The area. — 2

19. (a) Retention $\begin{cases}\text{Time}\\\text{Volume}\end{cases}$ — 2

(b) Chemical Reactivity. — 2 — 4

20. (i) (a) Must be miscible with the sample.
(b) Must not react with any component in the sample.
(c) Should give only one peak, which does not overlap any sample component peak.
(d) A retention time close to the sample component of interest. — Any 3, 1 mark each

(ii) Wt./volume % and peak height ratio or peak area ratio. — 2

(iii) 1 to 30% — 1 — 6

21. (i) 100 — 1
(ii) 700 — 1
(iii) 900 — 1 — 3

22. Nonane and Decane. — 2

23. (a) Lower the oven temperature
(b) Use less sample.
(c) Try a new gas flow rate (i.e. reduce carrier gas flow)
(d) Try a longer column. — Any 3, 1 mark each — 3

Total — 100

Appendix A
'A Nomogram for the Determination of the Retention Index'*

From the text (← **page 203**) we have seen that Kovats Index, I, is given by the equation:

$$I = 100\left[n\left(\frac{\log R_x - \log R_z}{\log R_{z+n} - \log R_z}\right)+z\right]$$

where R_x is the retention time of unknown substance X

R_z is the retention time of the normal alkane having z carbon atoms

R_{z+n} is the retention time of the normal alkane having $z + n$ carbon atoms

and n is the difference in the number of carbon atoms for the n alkanes.

The fact that to solve this equation both slide rule and logarithm tables are needed is a major disadvantage of the system. To overcome this disadvantage, Hupe† has devised a nomogram for the calculation, which involves only the drawing of either lines with a ruler, or a semicircle with a pair of compasses, followed by simple arithmetic.

Now look at the nomogram (→ **page 291**). On the ordinate, the time is marked in seconds from 10 to 3,600 (the number of seconds can be, if required, multiplied by one decimal value, giving the range then as 1 to 360 or 0·1 to 36 seconds).

* Reproduced by kind permission of Dr Ing. K. P. Hupe, 75, Karlsruhe 51, Lang Strasse 25 and Preston Technical Abstracts Co., 909 Pitner Ave., Evanston, Illinois.
† K. P. Hupe, *J. Gas Chromatog.* **3**, 12 (1965).

The actual range selected is determined solely by the lowest and highest retention values encountered.

Alternatively, the ordinate could be a length obtained from the chromatogram as retention time, e.g. mm, provided that these are related to the retention times by a constant factor. The range then would be 10 to 3,600 mm.

To use the nomogram, R_z is measured on the left ordinate and R_{z+n} on the right ordinate. The two points obtained are joined with a straight line.

(i) When R_x, the retention time of unknown substance X LIES BETWEEN those of the two reference substances, R_x is measured on the LEFT ordinate. Where the horizontal line drawn through this point intersects the line joining R_z and R_{z+n} (see Fig. A.1) a value I' is obtained which is related to Kovats Index I by the formula:

$$I = 100z + nI'$$

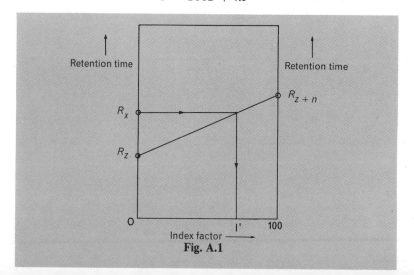

Fig. A.1

Let's see how it works in practice:

Using hexane and octane as reference alkanes, we have $R_z = 210$ mm, $R_{z+n} = 795$ mm, $R_x = 470$ mm, $n = 2$, $z = 6$. Then, from the nomogram $I' = 60.5$

$$\therefore \quad I = 100z + nI' = 100(6) + 2(60.5) = 600 + 121 = 721$$

(ii) When R_x, the retention time of the unknown substance is LESS than R_z, the line joining R_z and R_{z+n} is drawn as before. Then, using point R_z as centre, a semicircle of radius $(R_z - R_x)$ is drawn. A horizontal line is drawn through the point where the semicircle intersects the ordinate above R_x (See Fig. A.2). The value of I' obtained is negative, so the formula $I = 100z + nI'$ still applies.

Here is an example:

Using hexane and octane as reference alkanes, we have: $R_z = 210$ mm, $R_{z+n} = 795$ mm, $R_x = 178$ mm, $n = 2$, $z = 6$. Then from the nomogram, $I' = -12.25$

$$\therefore \quad I = 100(6) + 2(-12.25) = 600 - 24.5 = 575.5$$

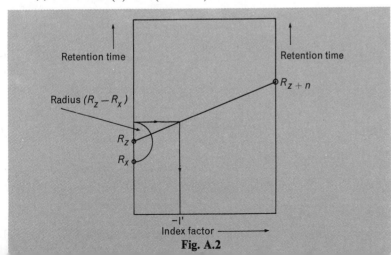

Fig. A.2

(iii) When R_x, the retention time of the unknown substance is MORE than R_{z+n}, the line joining R_z and R_{z+n} is drawn as before. Then R_x is measured on the RIGHT ordinate. Then, using point R_z as centre, a semicircle of radius $(R_x - R_{z+n})$ is drawn. A horizontal line is drawn through the point where the semicircle intersects the ordinate above R_z (Fig. A.3).

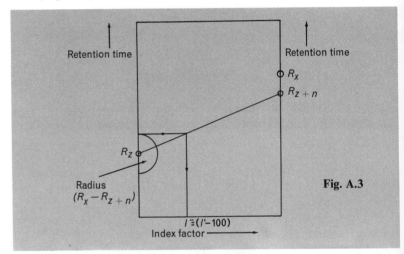

Fig. A.3

The value obtained, $I'' = (I' - 100)$,

From this equation I' is found, and on substituting it in the equation $I = 100z + nI'$, the retention index can be found.

Let's look at an example:

Using hexane and octane as reference alkanes, we have $R_z = 210$ mm, $R_{z+n} = 795$ mm, $R_x = 970$ mm, $n = 2$, $z = 6$.

Then, from the nomogram, $I'' = 14.5 = (I' - 100)$.

$$\therefore \quad I' = 114.5,$$

which gives $I = 100(6) + 2(114.5) = 600 + 229 = 829$.

A Nomogram for the Determination of the Retention Index

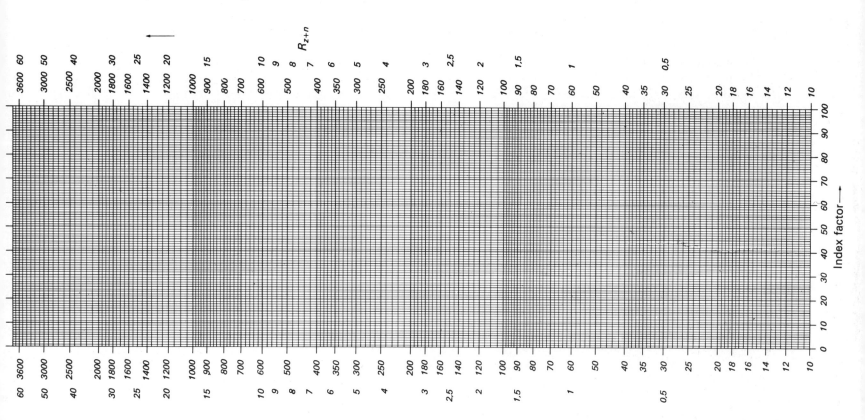

Appendix B
Pressure Couplings

The pressure couplings used to connect the column to the rest of the g.l.c. system incorporate the well-known principle of the compression joint in its simplest form, as shown in Figure B.1.

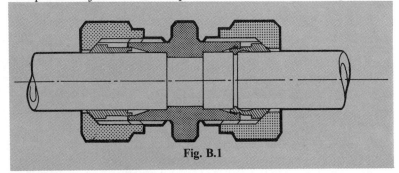

Fig. B.1

Couplings are available for use with plastic, glass, copper, aluminium or stainless steel heavy or thin wall tubing. A typical coupling is shown in Figure B.2.

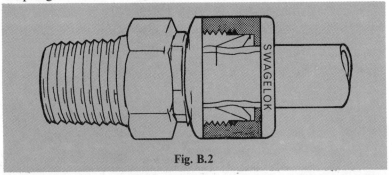

Fig. B.2

The tubing is supported over the entire length of the fitting and provides leak proof seals at three separate points. No torque is transmitted in tightening the fitting and the entire connection can be made in seconds.

Similar couplings to those made by Swagelok (marketed in the United Kingdom by Techmation Ltd.), are those made by Wade Couplings Ltd., Drallim Tube Couplings Ltd., Simplifix Couplings Ltd. and Hoke 'Gyrolok' Couplings (marketed in the United Kingdom by Hoke International Ltd.). The latter two companies supply fittings as small as $\frac{1}{16}$ in. o.d., suitable for use with capillary column tubing.

Appendix C

A Simple Gas Chromatograph

In order to test the effectiveness of the teaching of this programme, a simple, inexpensive gas chromatograph* was used by a group of sixth form students,† who prior to tackling the programme had not met or used the technique.

The detector fitted to this machine is of the katharometer type and the output signal was indicated by a galvanometer. The carrier gas used was nitrogen. The students found no difficulty with sample injection, and the controls enabled the carrier gas volume flow rate to be adjusted easily to any desired setting. Although the students found the procedure for setting up the galvanometer tedious, it was not difficult. In order to increase the usefulness of the machine, the column was wrapped with low voltage heating tape and insulated with fibre glass so that experiments could be carried out at elevated temperatures.

Here are the results of the experimental work using the standard instrument which incorporates a 4 ft general purpose column packed with silicone impregnated Celite.

Experiment 1. *To find the retention times of various gases*
Carrier gas pressure 5 psi; Volume injected in each case 0·5 ml; Temperature 15°C.

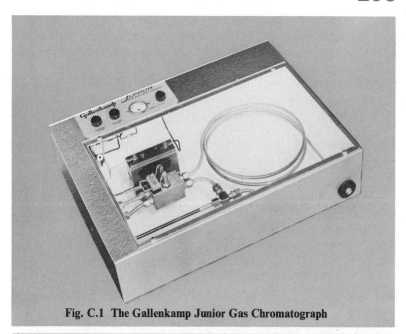

Fig. C.1 The Gallenkamp Junior Gas Chromatograph

* The Junior Gas Chromatograph and ancillary equipment were kindly loaned to the author by A. Gallenkamp & Co., (Northern) Ltd., Victoria House, Widnes, Lancs., for the purpose of these experiments.
† At A. J. Dawson Grammar School, Wingate, Co. Durham.

Gas	Galvo. reading (Max. Peak Height)	Retention time	
	divisions	min	sec
Ethane	3·2		35
Propylene	5·4		50
Butene-1	1·7	1	43
cis Butene-2	0·6		28
trans Butene-2	1·0	2	2
Butadiene	2·9	1	58

Experiment 2. *The identification of the components in a mixture*
The mixture consisted of methanol, methyl acetate and ethyl acetate. Carrier gas pressure 15 psi; Volume injected 5 µl in each case; Temperature 15°C.

(a) *Mixture alone*

Component	Max. Peak Height divisions	Retention time min	sec
A	0·8	2	6
B	2·3	4	4
C	1·0	10	10

(b) *Mixture + 2 drops of methyl acetate*

Component	Max. Peak Height divisions	Retention time min	sec
A	0·8	2	6
B	2·4	4	5
C	1·0	10	15

(c) *Mixture + 2 drops of ethyl acetate*

Component	Max. Peak Height divisions	Retention time min	sec
A	0·8	1	58
B	2·3	3	55
C	2·4	10	20

From these results component B was identified as methyl acetate, component C as ethyl acetate, hence Component A must be methanol.

Experiment 3. *To investigate the effect of column temperature on the separation of two substances having similar boiling points, namely, benzene, b.p. 80·1°C and cyclohexane, b.p. 80·8°C.*

(a) *At 14°C.* Carrier gas pressure 15 psi; Volume injected 5 µl of sample.

Component	Max. peak height divisions		Time min	sec
Cyclohexane	1·2	t_1	8	26
		t_2	9	18
		t_3	10	8
Benzene	1·5	t_4	11	25
		t_5	12	48
		t_6	13	55

The times taken were as follows for each peak:

t_1 and t_4—the time when the 'spot' on the galvanometer began to rise up the scale from the baseline position.

t_2 and t_5—the retention times of cyclohexane and benzene respectively,

t_3 and t_6—the time when the 'spot' on the galvanometer reached the baseline position once again.

(b) *At 65°C.* Carrier gas pressure 15 psi; Volume injected 5 µl.

Component	Max. Peak Height		Time	
	divisions		min	sec
Cyclohexane	4·7	t_1	1	30
		t_2	1	47
		t_3	2	24*
Benzene	6·4	t_4	1	43*
		t_5	2	10
		t_6	2	42

(* The 'spot' did not return to the baseline but at a position of 3·4 divisions, 1 minute 57 seconds from the injection point, began to rise up the scale once more. The values for t_3 and t_4 were obtained by extrapolation).

The following calculations were made from these results.

(i) *For the cyclohexane peak at 14°C*:

No. of theoretical plates,

$$n = 16 \left[\frac{(t_2)}{(t_3 - t_1)} \right]^2 = 16 \left(\frac{558}{102} \right)^2 = 480$$

As the length of the column was 4 feet then the Height Equivalent to a Theoretical Plate,

$$H = \frac{48}{480} = \frac{1}{10} = 0·1 \text{ in.}$$

(ii) *Calculation of the Peak Resolution, R_1 at 14°C*

$$R_1 = 2 \left[\frac{(t_5 - t_2)}{(t_3 - t_1) + (t_6 - t_4)} \right]$$

$$= 2 \left[\frac{(768 - 558)}{(608 - 506) + (835 - 685)} \right] = 2 \left[\frac{(210)}{(102 + 150)} \right]$$

$$= 2 \left[\frac{210}{252} \right] = 1·67$$

(iii) *Calculation of the Peak Resolution, R_2 at 65°C*

$$R_2 = 2 \left[\frac{(130 - 107)}{(144 - 90) + (162 - 103)} \right] = 2 \left[\frac{(23)}{(54 + 59)} \right] = 2 \left[\frac{23}{113} \right]$$

$$= 0·4$$

This experiment showed clearly, that an increase in column temperature of 51°C, although causing an increase in the peak heights, had reduced the peak resolution by one quarter, i.e. had caused a decrease in the efficiency of the column.

Other experiments, such as the investigation of the effect of varying the bridge current, the checking of laboratory reagents and samples prepared in their own laboratory for purity, and the identification of the impurities, were also carried out. The work is continuing and this report gives only the progress made at the time of the publication of this book.

Conclusions

It was easy to observe the maximum peak height on the galvanometer scale and to measure the retention time of each component with a stop-watch. Unfortunately, the rapidity of the response of the galvanometer made it difficult to plot the elution curve for any particular substance, but with the output connected to a suitable recorder (which is available, but adds considerably to the cost of the gas chromatograph), this disadvantage is eliminated. A recorder also provides a permanent chromatogram which could be filed for reference purposes. Typical chromatograms obtained using the recorder are shown in Figure C.2.

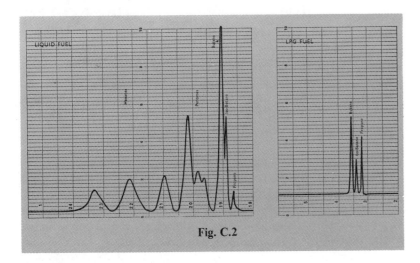

Fig. C.2

countered in their school laboratory. Coupled with the teaching programme, the knowledge gained of gas–liquid chromatography would be a distinct advantage for those who enter industry or University from school.

In the opinion of the group, this machine does serve as a most valuable introduction to the practical side of gas–liquid chromatography, and it works on the problems which they have en-

Appendix D

Syringes currently available

As well as the fixed needle syringe for liquid injections described in the text, models are also available for the injection of gases and solids into the gas–liquid chromatograph. Among the refinements available are models fitted with removable, interchangeable needles of various lengths, guided plungers and repeating adapters.

A 1 µl and a 10 µl syringe from the ranges of three manufacturers were chosen and some comparisons are tabulated (Table D.1). The leak rate (blow back) tests were done using ethylbenzene. The syringe was first wetted with sample, then the residual volume was fixed at 1 µl (1/10th of the syringe volume). The needle was then inserted through the septum with the carrier gas pressure at 10 psi, and held there for 10 seconds. If any of the sample had flowed past the plunger during this time, good reproducibility of injection would have been difficult. The reproducibility tests were done by injecting fixed amounts as indicated of ethylbenzene into a Perkin–Elmer F11 Flame Ionization Gas Chromatograph fitted with a column containing Apiezon L as stationary phase. The oven temperature was 166°C and the carrier gas pressure 10 psi. Four injections were made using the technique described in Part Seven, the peak heights were measured, their average obtained and the figure shown in the table is the maximum difference from that mean. None of these injections were made using a repeating adapter and no doubt by the use of such an aid these figures would improve. One of the disadvantages with all the 1 µl syringes is that the sample cannot be seen in the charged syringe, so it is essential to pump the plunger repeatedly to ensure the removal of the air bubbles before drawing the required volume of sample into the syringe. The terms 'removable' and 'replaceable' should not be confused, when used to describe the type of needle fitted to a syringe. In the case of the 1 µl and 5 µl (5 µl type B for the SGE) syringes, the needles are removable to allow the syringe to be cleaned, and refitting them is a delicate operation, but these needles are not replaceable with a new needle of similar size, nor are they interchangeable with needles from a similar syringe. In the larger sizes, e.g. Shandon–Terumo 10 µl to 100 µl syringes, the needles are replaceable and interchangeable as their external diameter is the same for all sizes.

As regards price, the syringes with replaceable needles offer distinct advantages over their fixed needle counterparts e.g. by buying an SGE syringe with guided plunger, removable (and replaceable) needle and repeating adapter for £11 (current price) one obtains a very durable, robust syringe which, despite the rather thick graduations and the initial rather unfamiliar feel of it (especially if one has been used to using a lighter syringe, c.f. 10 µl Hamilton), gives reproducibility good enough to suit most applications.

However, one cannot really point to a 'best-buy' for opinions will vary from person to person. The best advice is to try them all and choose the syringe which suits you and the application in which it is to be used.

Acknowledgement –I would like to thank Mr. J. Baxter of Scientific Glass Engineering Pty Ltd., for making available a wide range of their syringes, and also Mr. G. D. Hughes of Orme Scientific Ltd., who supplied the Hamilton and Shandon–Terumo syringes for examination and test purposes.

Table D.1: COMPARISON OF SYRINGES

Feature		Injection Size	Hamilton		S.G.E.		Shandon–Terumo		Notes
			1 µl	10 µl	1 µl	10 µl	1 µl	10 µl	(N.A. = not available)
Basic price (fixed needle)			N.A.	£8.25	N.A.	£6.00	N.A.	£6.00	Prices as at January 1972
Price with fixed needle and guided plunger or guide bar			N.A.	£11.50	N.A.	£8.00	N.A.	£7.00	
Price with removable needle			£16.00	N.A.	£12.00	£7.50	£12.00	£6.00	Does not necessarily mean that the needle is replaceable needle not supplied— charged extra.
Price with removable needle and guided plunger or guide bar			£19.25	N.A.	£12.00	£9.50	N.A.	£7.00	Needle not supplied— charged extra.
Replacement needles			N.A.*	N.A.	£1.50	25p to 75p	special order	87p	
			* Repairs cost 50% of the original price						
Repeating adapters			£2.75 †	£7.75 †	£1.50	£1.50	£1.25	£1.25	
			† Chaney repeating adapter. The Shandon Repr-o-jector can be used for both Hamilton and Shandon-Terumo syringes at £12.50.						
Graduations	Steps		0·01 µl	0·1 µl	0·02 µl	0·2 µl	0·01 µl	0·1 µl	Average thickness of 4 lines on barrel measured with a microscope fitted with a calibrated vernier eye piece.
	Lines $mm \times 10^{-2}$		42	47	65	85	21	34	
	Graduated length of barrel		5·5 cm	5·4 cm	5·7 cm	5·0 cm	5·0 cm	5·1 cm	
Reproducibility of Injection		0·4 µl	±2·3%	—	±2·5%	—	±1·3%	—	
		0·4 µl	—	—	—	±1·6%	—	—	
		0·5 µl	—	±1·5%	—	—	—	±2·1%	
		1 µl	—	±2·2%	—	±1·9%	—	±1·8%	
Robustness			fair	fair	good	very good	moderate	moderate	
Needle length			8 cm	5 cm	7·0 cm	5·0 cm	7·5 cm	7·5 cm	As fitted to syringes tested.
Syringe weight			12·9 g	7·8 g	19·8 g	25·9 g	14·9 g	12·6 g	

Appendix E
A Guide to Further Reading

Since A. J. P. Martin and A. T. James published their, by now, historic paper in 1952 on the separation and estimation of volatile fatty acids by gas–liquid partition chromatography[1] there has been a phenomenal growth of the art and in the amount of published literature relevant to it.

Most of the basic textbooks on gas chromatography were written on or about the 10th anniversary of the publication of Martin and James' work and whilst they are strong in theoretical detail, the practical information which they contain is limited.[2, 3] A more readable approach to the subject is provided by books on chromatography in general,[4, 5] analytical chemistry,[6] or on the use of physical methods in organic chemistry.[7] For those students who wish to cultivate a good practical knowledge, an excellent guide is available.[8]

For some analyses, specialist applications are necessary, for example, in programmed temperature work.[9] In this book, some knowledge of gas chromatography is assumed, and although each chapter includes some discussion of isothermal gas chromatography, the emphasis is on those aspects which are sensitive to temperature changes, and essential for an understanding of programmed temperature gas chromatography. For other specialist needs, books are available,[10, 11] and the catalogues of the larger booksellers will supplement this list.

As the art of gas chromatography developed it rapidly became clear that those interested in the subject should meet together to discuss the implications of the outstanding developments in the field. Periodic organisation of these symposia on an international basis started in 1956. The proceedings of them are published and form a source of useful reference material. The most recent of these publications are listed.[12–15]

At intervals, books of bibliographical references are published. These are very useful in the location of original literature references.[16, 17]

An essential book to anyone working in the field of gas chromatography is one listing retention data.[18] Although it is true that textbooks, reviews and journals all publish data, too, an organised reference list is invaluable.

In every art which is developing rapidly it is wise, from time to time, to take stock of the situation. This exercise is done for the chromatographer in a series of review publications. The early volumes of *Chromatographic Reviews* contained collected reprints of review articles which had appeared in the *Journal of Chromatography*. Later volumes also contained reviews which had not been published previously, and by 1966 had become an independent annual volume of review articles.[19] Sometimes a review monograph is published which reprints articles which have appeared earlier, but which, by being set out together cover a certain field of application.[20]

Even with all the help listed so far, it is still difficult for the practising gas chromatographer to keep up with the latest developments. By the time a paper appears in the usual abstracting journals (e.g. *Chemical Abstracts*) a long time may have elapsed since its first publication. It is here where a journal can help. There are several of interest, such as the *Journal of Chromatography*, the *Journal of Chromatographic Sciences*, *Chromatographia*, *Analytical Chemistry*, *Nature* and *Laboratory Practice*, but it is worthwhile noting that the *Journal of Chromatography* features a bibliography section covering all the techniques of chromatography, including gas–liquid chromatography. The title of the paper, the authors' names and the name of the publication in which it appeared are given. In 1968 the Gas Chromatography Discussion Group of the Hydrocarbon Research Group of the Institute of Petroleum commenced sponsorship of a compilation of abstracts in the field of Gas Chromatography. Membership of the Group is open to anyone, the fees are low, and as well as providing members with the abstracts the Group also sponsors International Symposia on Chromatographic topics.

Lastly, a very important source of the latest information is given by the *Abstracting Service* of an American company.[21] The service is so good that subscribers receive the abstracts weekly (on punched cards) within 2 to 4 weeks of the work being either published or presented. From January 1968, these abstracts have been available in a monthly bound volume which contains 235 to 240 entries.

References

1. A. J. P. Martin and A. T. James, *Biochem. J.,* **50**, 679 (1952).
2. A. B. Littlewood, *Gas Chromatography,* (*Principles, Techniques and Applications*) (2nd Edition), Academic Press, New York, 1971.
3. J. H. Purnell, *Gas Chromatography,* Wiley, New York, 1962.
4. D. Abbott and R. S. Andrews, *An Introduction to Chromatography,* Longmans, London, 1965.
5. R. Stock and C. B. F. Rice, *Chromatographic Methods,* Chapman and Hall, London 1963.
6. R. L. Pecsok and L. D. Shields, *Modern Methods of Chemical Analysis,* Wiley, New York, 1968.
7. F. L. J. Sixma and H. Wynberg, *A Manual of Physical Methods in Organic Chemistry,* Wiley, New York, 1964.
8. L. S. Ettre and A. Zlatkis, (Eds), *The Practice of Gas Chromatography,* Interscience, New York, 1967.
9. W. E. Harris and H. W. Habgood, *Programmed Temperature Gas Chromatography,* Wiley, New York, 1966.
10. H. P. Burchfield and E. E. Storrs, *Biomedical Applications of Gas Chromatography,* Academic Press, New York, 1962.
11. R. W. Moshier and R. E. Sievers, *Gas Chromatography of Metal Chelates,* Pergamon Press, London, 1965.
12. L. S. Ettre, (Ed) *Advances in Gas Chromatography,* Preston Technical Abstracts Co., Evanston, Illinois, 1967.
13. A. B. Littlewood, (Ed), *Gas Chromatography,* Institute of Petroleum, London, 1967.
14. L. R. Mattick and H. A. Szymanski, (Eds), *Lectures on Gas Chromatography —Agricultural and Biological Applications,* Plenum Press, New York, 1965.
15. L. Fowler, (Ed), *Gas Chromatography,* Academic Press, New York, 1963.
16. A. V. Signeur, *Guide to Gas Chromatography Literature Vol. 1,* Plenum Press, New York, 1964.
17. A. V. Signeur, *Guide to Gas Chromatography Literature Vol. 2,* Plenum Press, New York, 1967.
18. W. O. McReynolds, *Gas Chromatographic Retention Data,* Preston Technical Abstracts Co., Evanston, Illinois, 1966.
19. M. Lederer, (Ed), *Chromatographic Reviews Vol. 8,* Elsevier, Amsterdam, (1966).
20. M. A. Khan, (Ed), *The Fundamental Aspects of Gas Chromatography,* United Trade Press, London, 1962.
21. Preston Technical Abstracts Co., 909, Pitner Avenue, Evanston, Illinois.

Appendix F: SOLID SUPPORT EQUIVALENTS CHART

(Reproduced by kind permission of Phase Separations Ltd., Deeside Industrial Estate, Queensferry, Flintshire).

This chart has been compiled as a useful guide to solid support selection. In general, a separation of given efficiency may be found in any one of the excellent supports listed horizontally, but variations may be found particularly with highly silanised supports. Consequently, substitution of one support for another in the same horizontal line will generally be successful, but no guarantee is offered or implied.

Manufacturer	Phase separations	Johns-Manville	Applied Science	Analabs	Other	Description
Pink supports (Firebrick)	Phase Sep P Phase Sep P–AW Phase Sep P–HMDS Phase Sep P–AW–DCMS Phase Sep P–AW–BW Phase Sep P–1	Chromosorb P Chromosorb P–AW Chromosorb P–HMDS Chromosorb P–AW–DCMS no equivalent no equivalent	Gas–Chrom R Gas–Chrom RA no equivalent Gas–Chrom RZ Gas–Chrom RP no equivalent	Anakrom P Anakrom PA no equivalent no equivalent Anakrom S no equivalent	Sterchamol–German manufacture Diatomite S–U.K. manufacture See Phase Sep P	Plain Brickdust Acid washed brickdust Hexamethyl disilizane treated Acid washed and treated with dimethyldichlorosilane Acid washed–base washed Special liquid de-activation: temperature limited to 200°C.
White supports based on Celite 545†	Phase Sep CL* Phase Sep CL–AW* Phase Sep CL–AW–DCMS* Phase Sep CL–HMDS* Phase Sep CL–AW–BW* Phase Sep HC Phase Sep HC–1 Phase Sep HC–AW–DCMS Phase Sep CL–AW–B–DCMS	Chromosorb W Chromosorb W–AW Chromosorb W–AW–DCMS Chromosorb HMDS no equivalent no equivalent no equivalent no equivalent no equivalent	Gas–Chrom CL Gas–Chrom CLA Gas–Chrom CLZ Gas–Chrom CLH no equivalent no equivalent no equivalent no equivalent no equivalent	Celite 545 U Celite 545 A no equivalent no equivalent no equivalent no equivalent no equivalent no equivalent Celite 545 ABS	Celite† 545 available in many places Embacel is acid washed Celite 545 Diatomite C is also an alternative	Plain Celite 545 Acid washed Celite Acid washed and treated with dimethyldichlorosilane Hexamethyl disilazane treated Acid washed–base washed Untreated Celite 545 Special liquid de-activation: temperature limited to 200°C Acid washed and treated with dimethyl dichlorosilane Acid washed–base washed DC MS
White supports based on Celatom‡	Phase Sep N Phase Sep N–AW Phase Sep N–AW–DCMS Phase Sep N–AW–BW Phase Sep N–HMDS Phase Sep N–1 Phase Sep ABS	no equivalent no equivalent no equivalent no equivalent no equivalent no equivalent no equivalent	Gas–Chrom S Gas–Chrom A Gas–Chrom Z Gas–Chrom P no equivalent no equivalent no equivalent	Anakrom U Anakrom A Anakrom AS Anakrom AB no equivalent no equivalent Anakrom ABS	Diatomite CT no equivalents	Plain Celatom Acid washed Celatom Acid washed and treated with dimethyldichlorosilane Acid washed–base washed Hexamethyl disilazane treated Special liquid de-activation: temperature limited to 200°C Acid washed–base washed–treated with dimethyldichlorosilane
Special high quality supports	Universal B	no equivalent Chromosorb AW–DCMS no equivalent	no equivalent no equivalent Gas–Chrom Z	no equivalent no equivalent Anakrom ABS	no equivalents	Specially de-activated Celite: patented process. Acid washed and treated with dimethyldichlorosilane: Celite Acid washed and treated with dimethyldichlorosilane: Celatom
Super quality supports		High Performance Chromosorb AW–DCMS	Gas–Chrom Q	Anakrom SD	Aeropak 30 Diatoport S	Highly silanised versions of Celite or Celatom
High density supports	no equivalent	Chromosorb G	no equivalent	no equivalent	no equivalent	Available also as AW and AW–DCMS
Supports for preparative work	Phase Prep A no equivalent	no equivalent Chromosorb A	no equivalent no equivalent	AnaPrep no equivalent	no equivalents	Specially prepared from Celatom High density special support
PTFE based supports	Phase Sep T 6	Chromosorb T	Teflon 6	Tee Six	Haloport F Fluoropak 80	Various forms of PTFE
Porous polymer supports	Phase Pak P Phase Pak Q	no equivalent Chromosorb 102	no equivalent no equivalent	no equivalent no equivalent	Poropak P, Q, R, S, T & N (Waters Associates): Polypak 1 & 2	Cross linked polystyrene beads with varying pore sizes and polarities
Molecular sieve	Phase Sep Molecular Sieve 3A, 4A, 5A, 10X, 13X, etc.	no equivalent	Linde 5A and 13X	Linde 4A, 5A & 13X	no equivalents	High Alumina silicates with varying alkali metal substituents and pore sizes
Silica Gel/Alumina	Phase Sep Special grades for g.c.	no equivalent	Davidson D-actigel and Alcan	Davidson & Alcan	Various types available	Silica gel with standard porosity: gamma alumina (activated)

† CELITE: Trade Mark of Johns-Manville   * We strongly recommend the use of CELATOM based supports in   We must stress that this chart has been compiled for the benefit of all gas chromatography users to acquaint
‡ CELATOM: Trade Mark of Eagler Richer

INDEX

Acid, amino, 273
  carboxylic, 94, 109
    esters, 272
  fatty and resin, 272
Acrylonitrile, 84
Alcohol(s), 82, 84, 92, 94, 95, 104
Amines, 272
Analysis, qualitative, 155, 193, 277
  quantitative, 54, 58, 66, 68, 176, 208, 279
Apiezon L, 103, 113
Association, by hydrogen bonding, 82, 84, 92, 94, 109, 276
  self, 244

Base line, 58
  drift, 263, 270
Benzene, 8, 11, 17, 22, 25, 30, 83, 87, 92, 200, 294
  1-chloro, 83, 104
Benzonitrile, 99, 104
Boiling Point, *see* Vapour Pressure and,

Calibration graph, 214, 223, 280
Carbon tetrachloride, 97, 104
Carbon disulphide, 22, 25
Carbowax 20,000, 84, 103, 113
Carrier gas, *see* Phase, gaseous moving,
Celite, 115, 117, 276
Chloroform, 102, 104
Chromatogram, 58, 71
  interpretation of, 229 to 270
Chromatograph, simple gas, 293 to 295
Chromatography, 31,
Chromatography, gas-liquid, 10, 16, 22, 31, 39, 40, 52, 155, 274
Column, capillary, 271
  check for, 270
  conditioning, 131, 133, 139
  efficiency, 39, 257, 275
  glass, 31
    filling, 119, 121, 123, 125
  length and volatility, 170

Column, *cont.*
  metal, 31
    winding, 115, 133, 135, 137, 139
  overload, 250, 269
  packing, preparation of, 127, 271, 276
  and vapour pressure, 83, 92
Component identification, 271
Components, 30
Copper tubing, 31, 40
Curve, elution, 45, 275
Cyclohexane, 8, 11, 83, 92, 96, 294
Cyclohexanone, 200
cis-Cyclo-octene, 8, 16

cis-Decalin, 165
Detection, 58
Detectors, 58, 68, 70, 71, 155
  response factor, 63, 68
Di-nonyl phthalate, 84, 92, 113
Distillation, 5, 8

Ethanolamines, 273
Elements, electronegative, 82, 109, 276
Ether (*di*-ethyl), 5, 10, 11, 103
Ether(s), 94
Ethylbenzene, 200
Extraction, solvent, 22, 28

Firebrick, powdered, 40, 115, 117, 276

Hydrogen bonding, see Association

Iodine, 1, 10
Inflection tangents, 55, 66
Injection, sample, 143 to 152, 155, 271, 277
  syringe technique, 143 to 152, 277
Injection point, 58
Iron, 1, 10

Kieselguhr, 125, 127, 129, 131, 276
Kovats Index, *see* Retention Index, Kovats

Liquid Paraffin, 31, 40

Mercuric Bromide, 30
Mercury, 1, 25
Methanol, 103, 294
Methyl Acetate, 103, 294

Nonane, 165

iso-Octane, 11
n-Octane, 200
Olefines, 273

n-Paraffins, 207
Partition Coefficient, 16, 17, 22, 25, 28, 30, 31, 36, 39, 45, 63, 156, 274
Peak, air, 156, 278
    area, 55, 63, 68, 210, 221, 275, 277
        ratio, 210, 280
    height, 55, 190, 277
        ratio, 223, 280
    resolution, 55, 74, 275, 295
    separation, 78, 92, 237
    shape, 176
        asymmetrical, 250
    tailing, 207, 231
Phase, 25, 30, 40
    gaseous moving, 31, 36, 40, 42, 155
        and volume flow rate, 33, 39, 45, 51, 66
    liquid stationary, 31, 36, 40, 42, 77, 78, 83, 113, 115 to 139, 276
        and chemical activity, 103, 113
        concentration on support material, 186, 276
Plate(s), effective number of, 173, 184, 277
    height equivalent to a theoretical (H), 66, 117, 275
    number of theoretical, 66, 275
    theory, 39, 40
Polarity, 83, 97, 104, 276
    group, 97, 276
Polyethylene glycol, *see* Carbowax 20,000
    adipate, 113
Pressure couplings, 292
Pressure ratio, 229, 235, 281
Programming, flow, 266
    linear temperature, 263, 270
Propane, 83, 90

Retention index, Kovats, 203, 278, 289

Retention index, *cont.*
    and stationary phase, 200
Retention time, 33, 39, 45, 51, 58, 63, 66, 77, 198, 275
    and column temperature, 51, 186, 235
    and concentration of stationary phase, 51
    equation for, 45
    factors influencing (summary), 241, 275, 281
    relative, 172, 206, 277
    and solubility, 77
    and volatility, 174
    and volume flow rate of the carrier gas, 51
    volume, 39, 45, 51, 58, 63, 66, 77, 275

Sample mixture, unequal concentrations, 234
Sensitivity, 77
Separation Factor, *see* Peak Resolution
Silicone Fluid, 113
Solvent, 115, 117, 131, 133, 135, 137, 139, 225
    mixture, 25
    and solute, 39
Squalane, 84, 113, 165
Standard, internal, 206, 208, 278
    international list, 195
    rules for, 207
    solution of, 221
Syringes, types of, 297

Tailing, *see* Peak, tailing
Temperature, oven, choice of, 170, 176, 178
    and partition coefficient, 156, 235
    and solubility, 39
Tetralin, 5, 11
Toluene, 8

Vapour pressure, and boiling point, 1, 10, 11, 16, 103, 156, 174, 274
    equilibrium, 36, 42
    of stationary phase, 83, 277
Volatility, see Vapour pressure.

Water, 1, 10, 17, 22, 25, 30

m-Xylene, 11
p-Xylene, 8, 16, 200

The Hamilton range
of precision syringes
and related equipment,
are the finest
and most precise
instruments
ever produced for
chromatographic
applications.

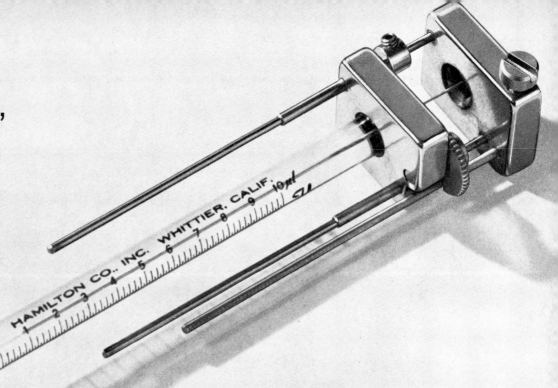

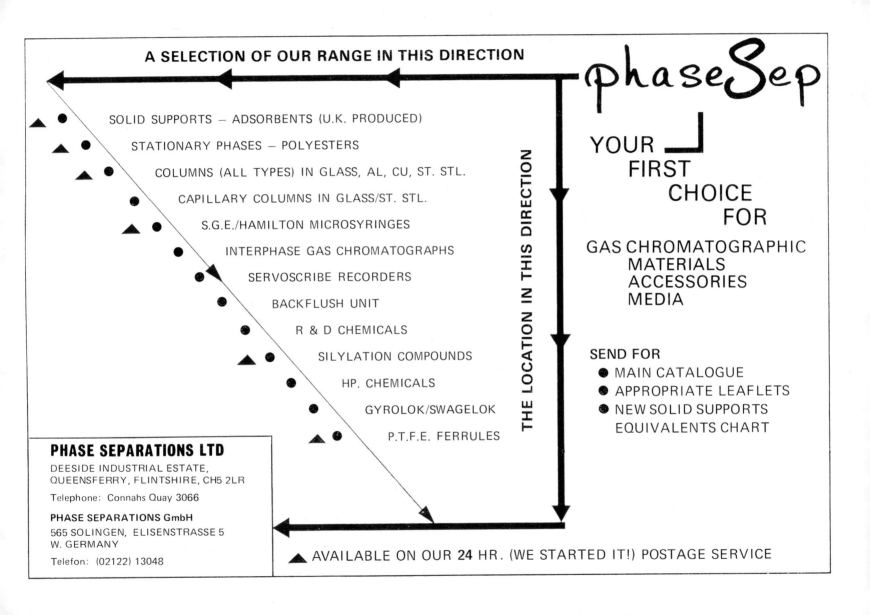

TRACOR GC'S CAN DETECT MORE.

(MORE DETECTORS)

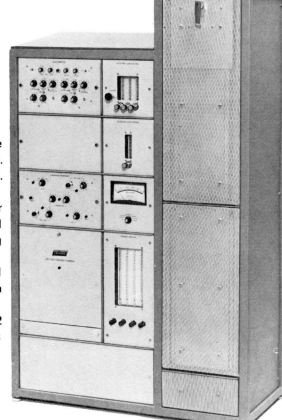

Equip your TRACOR gc from a book full of detectors. Three of them — the flame photometric, the Ni^{63} electron capture and the Coulson electrolytic conductivity detector — are TRACOR exclusives. And detectors are just the beginning: the Model 550 low-cost gc (left) and the advanced 222 G.C. research gc offer a total range of capabilities and features available nowhere else!

The 222 G.C. lets the researcher choose from 13 separate detector systems. Run any four simultaneously. Quadruple all-glass U columns, precision rotameters for exact gas flow control, and an extraordinary inlet system assure top performance and operating convenience.

With the Model 550, you can get automatic sampling and designed-in compatibility with all TRACOR detectors. It mates with mass spectrometers, and features dual all-glass columns. From $2500.

For more information about TRACOR's Model 550 and 222 G.C., contact Ken Mahler, Sales Manager, Analytical Instruments Division, 6500 Tracor Lane, Austin, Texas 78721.

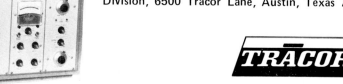

TRACOR
™

You must admit they've got a point

SGE decided a long time ago that, as far as the chromatographer was concerned, the best was not good enough. So they started on a continuing development programme to set new standards in precision microlitre syringes. It was not only a question of precise measurements but also of increasing the working life, the durability, and the flexibility that mattered. It was important that the downtime of a syringe should be the bare minimum.

All this holds true today, and you will find in all SGE syringes many results of their continuing development programme.

S.G.E. syringes are available from either of our offices or from distributors and leading instrument manufacturers throughout the world.

* Specially strengthened guided plungers are almost impossible to bend.

* Removable needle syringes can be used with low melting point solids.

* Double sheathed needles available for extra strength.

* Removable needle syringes permit replacement of damaged needles with no increase in dead volume and at low cost.

* Removable needles with three bore sizes for viscous materials.

* High pressure syringes which can withstand 10.000 p.s.i.

  If you would like to know more about our syringes please send for a set of our information sheets.

Scientific Glass Engineering Pty Ltd.
Head Office: P.O. Box 185, North Melbourne, Australia 3051.
European Office: 657 North Circular Road, London NW2 7AY.